BRUNEL
The Great Engineer

BRUNEL
The Great Engineer

TIM BRYAN

Ian Allan
PUBLISHING

Contents

Half title: Lithographed portrait of Brunel from a painting by J. C. Horsley. *Swindon Museum Service*

Title page: The Royal Albert Bridge over the River Tamar. Photographed in 1939. *British Railways*

Above: The west portal of Box Tunnel. A lithograph by J. C. Bourne. *Swindon Museum Service*

For Ann, with love.

First published 1999

ISBN 0 7110 2686 6

Published by Ian Allan Publishing

an imprint of Ian Allan Publishing Ltd, Terminal House, Shepperton, Surrey TW17 8AS.

Printed by Ian Allan Printing Ltd, Riverdene Business Park, Hersham, Surrey KT12 4RG.

Code: 9911/A3

Author's Preface

The interest in, and enthusiasm for, the story of Isambard Kingdom Brunel, perhaps one of the most famous and charismatic engineers of the Victorian era, seems undiminished; as evidenced by numerous books, both the public and the media are still drawn to a man who in his own lifetime captured the imagination and sometimes indignation of the people whilst producing bridges, tunnels, stations and steamships, many of which still survive today. With the promotion of Brunel's main line from Bristol to London as a potential World Heritage Site, it also seems an apt occasion to revisit the life of such an important figure.

Reviewing Brunel's life, one is struck by the sheer volume of work in which he involved himself; quite apart from the strain it must have put on his constitution, it presents a problem for the biographer, who cannot easily describe the progress of projects which were carried out simultaneously. As a result, although a broadly chronological approach has been taken, the book is divided into six main sections, each telling part of the overall story. I have, however, included a chronology of the main events of Brunel's life at the end of the book, to link the whole story together. A conscious decision was also made not to include absolutely every project Brunel worked on, so the reader is recommended to look at some of the other sources listed at the end of the book for further information.

As a Bristolian, and someone who has been professionally involved with the Great Western Railway and its history for well over a decade, I naturally wanted to write about this great man. There have already been a good number of excellent biographical studies of Brunel, many now out of print, and this book is intended to complement those writers who, like myself, have been drawn to retell the life of a man who, though small in stature, was big on ideas.

Tim Bryan
Swindon, September 1999

Acknowledgements

Although writing is a somewhat solitary business, no author can complete a work such as this without the assistance and patience of many others. I would particularly like to thank the staffs of the various libraries and institutions who have helped locate material or illustrations for this book. These include Bristol Reference Library, Swindon Reference Library, Bath University Library, Bristol University Library, Brunel University Library, Bristol Museum & Art Gallery, The National Portrait Gallery and the National Railway Museum at York.

Thanks are due to the various organisations which made illustrations available for reproduction in this book; any pictures not individually credited are from the author's collection apart from engravings of the Thames Tunnel, which are from Beamish's *Memoir of The Life of Sir Marc Isambard Brunel*.

There are also a number of individuals who gave time, advice and help during the writing of this book: Matthew Tanner, Curator of the SS *Great Britain* Project allowed me access to the Conservation Plan for the ship and dockyard which contained much valuable information, and also spared the time to show me around the recently discovered remains of the Great Western Steamship Co's works, next to the dock where the ship was built. I am also indebted to Gwyn Richards of Playback Productions, who kindly loaned me his Robert Howlett portrait of Brunel. Thanks are also due to Ian Nulty of Railtrack Great Western, Keith Falconer at the National Monuments Record Centre, Swindon, Beverley Cole and Ed Bartholemew at the National Railway Museum, Mike Pascoe at the Clifton Suspension Bridge Visitor Centre and Andy King at the Bristol Industrial Museum.

As usual I should like to thank the team at Swindon Borough Council for their support, Julia Holberry, Head of Cultural Services, Robert Dickinson, Principal Curator, and the whole team at the GWR Museum, including Julie Adams, David Berry, Elizabeth Day, Marion Flanagan, Alf Neate, Reg Palk, Marion Robins and Christine Warren.

No book would be written without the help of family and friends, and as usual I should like to thank my wife Ann for putting up with my absence while I researched and wrote the book, and for reading and correcting the text. Thanks are also due to my mother and sisters, for their help with childcare when I was visiting Bristol to research the book.

Left: An early portrait of Brunel, by his brother-in-law John Horsley, standing next to plans of the Great Western Railway.
National Railway Museum

Introduction
Early Life

The terraced houses in Britain Street, Portsea, a district of the naval city of Portsmouth, have long gone, and there is only a small plaque to signify the place where Isambard Kingdom Brunel, one of the most important engineers of the Victorian era, was born. The young Isambard was the third child and only son of Marc Isambard and Sophia Kingdom Brunel, and was born on 6 April 1806.

Marc Isambard Brunel had been born in 1769 in Hacqueville, a small town in Normandy. The path mapped out by his father Jean Charles had been for Marc to become either a lawyer or priest since, as the second son, he stood little chance of inheriting the family farm. With this in mind he received a classical education for which he was completely unsuited, preferring to spend time at a local carpenter and wheelwright's workshop. In 1780 he was sent to a Rouen seminary, where his interest in mathematics, geometry and drawing were recognised, and his complete lack of enthusiasm and vocation for the priesthood realised.

After only two years Marc Isambard left the seminary; despite opposition from his father, he was taken in by a distant relative, Captain François Carpentier, a retired naval officer, and trained to become a naval cadet. After almost four years of study, living with the Carpentier family, he joined his first ship, and was sent to the West Indies. Returning to Rouen in 1792, Marc Isambard found France much changed; the French Revolution had broken out three years earlier, and the country still seethed with revolutionary fervour. Visiting Paris with Carpentier, the two avowed Royalists had a lucky escape and fled back to Rouen hours before barricades were erected around the city. While staying with his adopted family once more, Marc Isambard met Sophia Kingdom, the daughter of Plymouth naval contractor William Kingdom. The 16-year-old Sophia had been sent to stay with the Carpentier family in order to improve her French, and soon fell for the dashing Royalist. Marc Isambard made little secret of his views, and as the Republican Terror spread throughout the country, it became apparent that, if he did not flee, his fate would lie with the guillotine.

Having become engaged to Sophia, Marc Isambard left France in July 1793, using a passport obtained through the American vice-consul in Le Havre, again through the offices of his friend Carpentier. The destination for his flight was the United States, and he remained there for the next six years, becoming an American citizen in 1796. Towards the end of his stay, Marc Isambard was introduced to Alexander Hamilton, a close friend of George Washington. During a

Above: Marc Isambard Brunel: portrait painted in 1835 by Samuel Drummond. *National Portrait Gallery*

dinner, Brunel was introduced to another French émigré, and during a discussion on the merits of the Royal Navy the matter of the manufacture of blocks for ships was mentioned. Marc Isambard felt sure that, with better equipment, the manufacture of these vital components of warships could be speeded up. As a Frenchman, Marc Isambard would, not surprisingly, be viewed with some suspicion by the British authorities, but a letter of introduction from Hamilton to Earl Spencer, First Lord of the Admiralty, helped smooth his passage when he arrived in England in the spring of 1799.

There is little doubt that another reason for Marc Isambard's wish to travel to England was Sophia Kingdom. In the six years they had been separated things had not been easy for her; after her suitor had fled to the United States she had been imprisoned, not returning to England until 1795. There appears to have been correspondence between the two while Marc Isambard was in America and, arriving in England in March 1799, he went straight to London, where

Sophia was staying with her brother. Despite the six-year separation, their engagement was confirmed, and in November they were married.

However good Marc Isambard's proposals for block-making equipment were, they could be of no use to him unless they were used commercially, and Fox & Taylor, the Southampton company which had held the monopoly on production, rejected his ideas out of hand. Although a skilled draughtsman, Brunel needed to be able to demonstrate his ideas using three-dimensional models, and a meeting with Henry Maudslay, a London craftsman, enabled him to argue his case with the Navy, since Fox & Taylor's contract was due to run out. Sir Samuel Bentham, Inspector-General of Navy Works, was sufficiently impressed to accept Brunel's proposals, and Marc, Sophia and their new daughter Sophia moved to Portsmouth, in order that Marc could supervise the construction of a block-making works, using machines built by Maudslay. The construction and commissioning of the new factory took six years, during which Isambard and his sister Emma were born.

By 1807 Marc Isambard wished to expand his activities, and he went into partnership to set up a sawmill and veneer works at Battersea, in London. This was an attempt to earn some money for his growing family since, although the Navy was more than happy with the success of the block-making plant, Marc Isambard was having great difficulty in extracting payment. Since the Portsmouth operation was now working well, a new, grander home for the family was obtained at Lindsey Row, Chelsea (now part of Cheyne Walk). During this time Marc Isambard also added a boot-

making plant close to the sawmill. However, the family finances were always shaky, and in 1814 matters took a turn for the worse. The first blow was the end of the Napoleonic War, which meant that there was no longer a need for large quantities of soldiers' boots; in August a further disaster occurred when the Battersea sawmill burned down. It was found that his business partner had been less than honest, and although Marc was able to finance the rebuilding, only a sympathetic banker kept the concern afloat.

From early on, it was clear that Marc Isambard expected much of his son. The 'apple of his eye' was expected to take over his mantle and become an engineer too, and it was recorded that, before sending his son to Dr Morell's boarding school in Hove, he had already taught him drawing and geometry. These skills seemed to serve him well, and in a letter written to his mother in 1819 he noted that he was undertaking a survey of the town of Hove, and drawing some of the larger houses there. A year later, aged 14, he was sent to France to finish his schooling at the College of Caen in Normandy, moving from there to the Lycée Henri-Quatre in Paris. Some clue to the origin of Isambard's eye for detail can also be found in the fact that his father also arranged for him to have a period of apprenticeship under Louis Breguet, the famous maker of chronometers, clocks and scientific instruments. During his stay with the craftsman, it appears that Isambard also studied at the Institution de M Massin; a recently-auctioned school report for Isambard reveals that his general conduct was 'irreproachable', and details prizes for French, drawing and mathematics.

During Isambard's stay in France, his father's financial affairs came to a catastrophic halt when his bankers failed, and Marc was declared insolvent. On 14 May 1821 he was sent to the King's Bench, a debtors' prison. The loyal Sophia went with him, and there they remained until July, the government refusing to pay Marc Isambard for the work he had done. His release was only made possible when he threatened to emigrate to Russia to take up an invitation made by Tsar Alexander some time earlier, when visiting the block-making works. This veiled threat was enough to force the government to grant Marc Isambard £5,000, enabling him to pay off his debts, but leaving him without any other income. Marc Isambard was never again to risk all in large ventures such as the sawmill, and began to work as a consulting engineer in a small office at 29 Poultry, in the City of London. All through what the Brunel family were later to call the 'Misfortune', Isambard had remained in France, supported by friends and benefactors, and did not return to England until August 1822.

When Isambard did return, aged 16, he already had both an excellent education and the benefits of working for the master craftsman Louis Breguet. Having suffered so much through the 'Misfortune', Sophia must have had grave misgivings about her son's joining the 'family business'.

Above: The plaque marking the spot in Portsea where Isambard Kingdom Brunel was born. *Swindon Museum Service*

However, Isambard seems to have needed little encouragement to work with his father, who was now engaged on a variety of projects, including two suspension bridges on the Ile de Bourbon, and similar structures over the Serpentine and the River Thames at Kingston, as well as designs for docks in East London, a sawmill in British Guyana and a variety of other jobs, large and small. Isambard also gained vital experience of a more practical nature working in Maudslay's workshops — the start of a working partnership which would continue for many years.

In the midst of this busy office, both father and son expended much effort on another project which would eventually come to nothing. The 'Gaz' or 'Differential Power' engine occupied much of Isambard's free time and initially promised much; the engine worked on the principle that gases could be liquefied by cooling and that, by raising the temperature of the gas, the resultant increase in pressure could be used to drive machinery.

Ultimately, the Gaz Engine did not prove to be the successor to steam power as many, including the Brunels, had hoped; the technology to manufacture machinery able to cope with pressures of up to 1,000psi did not exist, and it is a testament to the engineering skill of both men that there were no serious accidents during their unsuccessful experiments. From the beginning of 1823, however, they began work on the project which would occupy them for the next 20 years, and nearly cost Isambard his life — the Thames Tunnel.

Below: The Rotherhithe end of the Thames Tunnel: the large ramped ends to the tunnel giving access to vehicles originally planned by the Brunels were never completed due to lack of funds. *Illustrated London News*

ROTHERHITHE ENTRANCE TO THE TUNNELL.

Chapter One

Tunnels, Bridges and Docks

Two unsuccessful attempts had already been made to drive a tunnel under the River Thames before Marc Isambard Brunel was appointed as the Engineer to the Thames Tunnel Company in February 1824. By the early years of the 19th century there was a need for some additional crossing of the river in London, particularly in the vicinity of London Bridge and the East End, where the docks were expanding rapidly. There was severe congestion, not only on the roads of the city itself, but also on the river, where over 350 ferries plied their trade. The second attempt to drive a tunnel through the Thames mud had been made by Cornish engineers Robert Vazie and Richard Trevithick in 1807, but despite the construction of a 100yd pilot tunnel which ran from Wapping to Rotherhithe, the project was abandoned when the workings flooded. The tunnellers had struck quicksand, and the opinion of many scientists was that, despite its importance, the tunnel was an almost impossible proposition.

What was needed was the technology to allow tunnelling through such treacherous ground; Marc Isambard gave considerable thought to the problem, and in January 1818 filed a patent for 'Forming Drifts and Tunnels Underground'. He had been inspired by nature when working at Chatham Dockyard. Examining a piece of timber removed from the keel of a ship, he noticed that it was riddled with a series of holes made by the mollusc *teredo navalis*, which reputedly had sunk more ships than had enemy cannon. This nine-inch-long shipworm could bore through the hardest oak, and provided the inspiration for the elder Brunel to design an ingenious tunnelling shield. Before explaining these ideas, both at the Institute of Civil Engineers and in an article in *The Mechanics Magazine* in September 1823, Marc and Isambard had spent a good deal of time planning the details of the projected new tunnel, so much so that reasonably accurate estimates of expenditure and projected income from any new tunnel had been calculated.

On the day after the meeting with the Institute of Civil Engineers, 18 February 1824, a public meeting was held at the City of London Tavern, and a new body, the Thames Tunnel Company, was set up to build the tunnel. Enthusiasm for the project was high, and within weeks nearly £180,000 had been pledged by shareholders. The elder Brunel still had much work to do, lobbying various businessmen and politicians including the Duke of Wellington; a contemporary writer quoted the Iron Duke, who noted that there 'was no work upon which the public interest of foreign nations had been more excited than it

had been on this tunnel'. Within six months, the Bill giving the company powers to construct the tunnel had been given Parliamentary assent. At the first General Meeting, held on 20 July 1824, the appointment of Marc Isambard Brunel as Engineer to the Company was confirmed, at a salary of £1,000 per year for three years only, since it was hoped to complete the work within this time. The directors also agreed to pay for the use of the tunnelling shield patent, with £5,000 payable when the structure of the tunnel was complete, and a further £5,000 when the new company started to receive income from tolls levied on users of the tunnel.

The tunnel was to be sited downstream from London Bridge, and was to link Rotherhithe and Wapping in a 'populous and highly commercial neighbourhood', as one writer described it. The line of Brunel's tunnel was about three-quarters of a mile from the route chosen by Trevithick and Vazie; an immediate task was to carry out a geological survey to ensure that the misfortunes suffered by the Cornish engineers were not repeated. Isambard was heavily involved

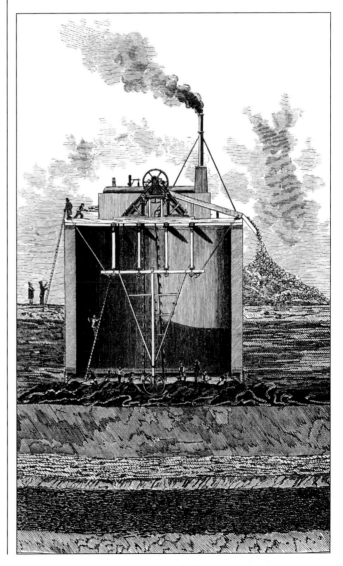

Left: The interior of the Tunnel after completion.
Swindon Museum Service

Right: The sinking of the main shaft for the Thames Tunnel at the Rotherhithe end. The steam engine was not ready when work started to sink the shaft.
From Beamish: Life of Marc Isambard Brunel *(Author's Collection)*

with this process and, as well as shafts sunk on either bank, a series of parallel borings was taken across the river itself. A layer of impervious blue clay was discovered, and it was hoped that this would act as a roof to the workings underneath. On 2 March 1825, to the sound of the bells of St Mary's Church in Rotherhithe, work began on what was described as a 'national enterprise'. Opening festivities over, work started on the sinking of a vertical access shaft; this was built by the construction of a 40ft-high brick tower, which was built on top of an iron ring with a cutting edge fixed to it. Resting on a ring of wooden piles, the tower was reinforced with metal tie-rods, and the whole structure was

then rendered with Roman cement. Once built, the wooden piles were taken away, and workmen then excavated soil around the base of the tower, which was removed by windlass, since the steam-powered bucket elevator, designed by Marc Isambard and due to be built by the London company of Maudslay, Son & Field, did not arrive in time. As soil was removed, the whole structure slowly sank under its own weight; the operation attracted a good deal of attention, and numerous dignitaries, including the Duke of Wellington and Sir Robert Peel, came to witness the sight of this great masonry tower, which weighed over 100 tons, slowly disappearing into the ground.

Left: Marc Isambard Brunel's patent tunnelling shield.
From Beamish: Life of Marc Isambard Brunel *(Author's Collection)*

Above: Using the diving bell to check damage to the tunnel. Brunel was reported to have taken his mother down in this dangerous equipment to view the workings!
From Beamish: Life of Marc Isambard Brunel *(Author's Collection)*

By early June it had sunk enough to allow the whole structure to be underpinned by a further 25ft of brickwork, which would allow the tunnelling shield to be put into position. Considerable trouble was experienced with the ingress of water during this operation — a foretaste of troubles to come — and a sump was excavated at the bottom of the shaft to allow water to be drained out of the workings and then pumped to the surface. In November 1825 work could finally begin on the tunnel itself; Marc Isambard Brunel's tunnelling shield was an ingenious and safe method of excavating, consisting of an 80-ton cast-iron frame, which was divided into 12 cells, each having three working areas, which allowed 36 miners to work at one time. Resting on cast-iron feet, and having similar fixings called 'staves' at roof level, the shield allowed 800sq ft of ground to be excavated at any one time. In front of the shield was a layer of oak planks, known as a 'poling board', held against the earth by jacks. Workmen removed one plank at a time, excavating behind it and then replacing the board and retightening the jack. When this process had taken place across the whole face of the excavation, the whole shield could be inched slowly forward. Behind the miners followed a wheeled trolley, from which workmen both constructed the brickwork of the tunnel itself and removed the spoil from the shield.

The hope that the tunnel would be completed within three years soon proved hopelessly over-optimistic. Although work had started in earnest on 28 November 1825, by May the following year the tunnel had only advanced 100ft. Conditions in the workings were far from ideal; the crown of the workings was only 14ft from the riverbed, and since the River Thames was at this time an open sewer, the men were often tunnelling through soil which was stinking and putrid. Many of the labourers employed to dig in the gloomy darkness of the tunnel were illiterate and unskilled and needed much supervision; bearing in mind the conditions, it is not surprising that drunkenness and ill-health were commonplace. In March 1826 Isambard was injured when a lump of timber fell on him in the workings, leaving him incapacitated for three weeks. Marc Isambard, at 55 years old, was in no condition to spend long periods in this hellish environment, and fell ill with pleurisy in April. Not long afterwards, the Resident Engineer, William Armstrong, was also taken ill, having had to put in three six-day shifts of almost continuous work while Isambard was injured. Although he returned for a spell, he eventually resigned in August.

The younger Brunel was thus left in sole charge, although without additional pay or official appointment as Resident Engineer. Marc Isambard was reluctant to confirm his appointment for fear of any accusations of nepotism, and thus appointed three assistants, Richard Beamish, William Gravatt and Francis Riley, to remove some of the strain from his son, who spent long periods underground, supervising the work. By January 1827 the directors of the company had presumably decided that the younger Brunel had proved his worth in the tunnel, since he was then formally appointed as Resident Engineer. As the year progressed, the danger of flooding steadily increased, as the workings appeared to be very close to the bed of the river itself. If this were not dangerous enough, the directors, impatient at what they saw as slow progress on the project, opened the workings to the public, in an effort to generate income. Quite why anyone would have wished to pay a shilling to descend into the dark, disease-ridden tunnel is questionable, but the presence of such visitors filled both Marc and Isambard with dread at the thought of what might happen if the river were to burst into the workings.

Instead of the clay originally encountered, the miners were increasingly meeting sand, gravel and water, as well as refuse and debris such as bones and china from the riverbed. Pumps were constantly required to extract water from the workings, and in March Isambard recorded in his diary that the area being excavated was 'extremely tender'. Matters were made worse by the fact that the line of the tunnel passed through an area of the riverbed which had been dredged for gravel, so that the water was extremely close overhead. After a number of false alarms, on 18 May the river finally surged into the tunnel, and completely flooded the workings. At the time, over 150 men were at work, so one can imagine the panic which overtook them when the wave of filthy water swept down the tunnel, extinguishing all the lights. Fortunately there were no fatalities or serious injuries, Isambard distinguishing himself by bravely climbing back down the shaft to rescue a trapped workman. Descending in a primitive diving bell, Isambard later inspected the hole in the riverbed, and a large raft was sunk over the breach, on to which was laid around 150 tons of clay in sacks. He made a number of hair-raising descents in the diving bell, including one when he again had to act quickly to rescue someone, this time one of his assistants, Pinckney, who had slipped out of the craft.

The task of plugging the breach made by the river, pumping out the water and then removing the silt and mud which had filled the tunnel took six long months. On 27 June the adventurous Isambard took a party, of those who had been working on the shield when the water rushed in, to view the workings. With only candles to light their way, they rowed up the western side of the tunnel in a punt, until they found a bank of silt blocking their way. Two days later the operation was repeated in the eastern section; what was apparent from both perilous journeys was that much work needed to be done to repair the tunnelling shield and remove the tons of debris and flotsam which now filled the workings. As water was pumped out of the tunnel, conditions for those working in it became worse; the foul air caused the workmen to feel giddy and sick, and a black deposit formed around their mouths and nostrils. Throughout all this work, Isambard spent many hours, either in the shaft or in the equally dangerous diving bell, inspecting the breach in the riverbed, which proved difficult to plug, breaking through on a number of occasions.

In October the tunnelling shield was finally repaired, and work could recommence; no sooner had this happened, than

Below: After the first influx of water, it was necessary to take a dangerous trip down the tunnel to see the extent of the damage, once the water had subsided. Brunel is said to be the figure in the narrowest part of the tunnel. *Author's Collection*

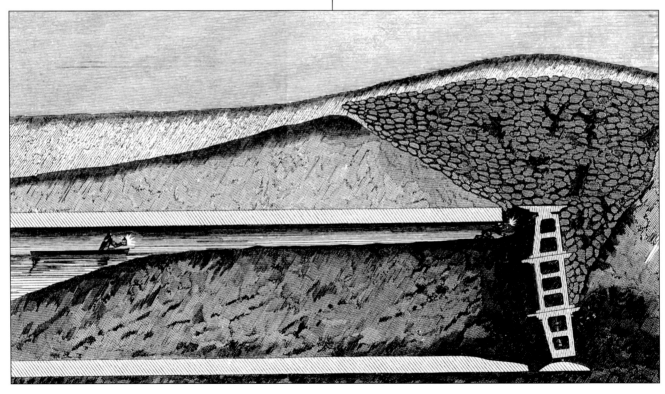

CEREMONY OF OPENING THE TUNNEL.

Isambard was once again injured when he fell into a water tank near the entrance to the shaft. He was lucky to escape with bruises and a severe shaking, but his injuries were enough to keep him away from work for almost three weeks. With work restarted on the project, it was decided in November 1828 to hold a celebratory dinner for the directors of the Tunnel Company, and guests. Isambard was the host of this rather surreal event, his father deliberately absent in order that the role of his successor be made plain. The sides of the tunnel arch were draped in crimson fabric and the band of the Coldstream Guards serenaded the assembled diners. As well as the 50 or so guests in the western tunnel, 120 workmen were in the next archway, enjoying a less elaborate meal of beef and beer. After the toasts, the leader of the miners joined Brunel and his guests to present him with a ceremonial pick-axe and spade. The young Isambard clearly enjoyed the respect of the workforce (despite the pay cuts and redundancies he had been forced to implement), due largely to the long hours he spent working with the men, in terrible conditions.

Early in 1829 there were ominous signs that all was not well. In addition to numerous water leaks, there had been reports of debris seeping down from the river bed. In the early hours of Saturday 12 January disaster struck again; water rushed in and surged down the tunnel, carrying all before it, including Isambard and all the workmen then working at the face. Before being overwhelmed by the torrent, Isambard had attempted to help miners working on the shield to escape. All lights were blown out, and everyone including Brunel was plunged into the flood. Swept along by the filthy water, Isambard was plucked unconscious from the foot of the main shaft by Richard Beamish, who had just arrived on the scene. Six men were not so lucky, and drowned in the darkness of the tunnel. Regaining consciousness, Brunel refused to leave the scene and, lying on

Above: The opening ceremony at the Tunnel in 1843. *Illustrated London News*

Below: Marc Isambard, pictured by the *Illustrated London News* at the opening of the Tunnel. *Illustrated London News*

a mattress situated on the deck of a barge, directed the operation to send down the diving bell to examine the breach. It was not until Monday 14 January that he finally retired to bed, where, tended by his mother and various doctors, he slowly began to recover. A few days later, the directors of the Thames Tunnel Company passed a resolution which was widely reported in the press. In it they praised Brunel's 'intrepid courage and presence of mind', and the way in which he had 'put his own life in more imminent hazard to save the lives of the men under his immediate care'. Although Isambard would now play little direct part in the project, repair work began once again, continuing until the summer when, to the dismay of his father, money ran out, and work on the tunnel was halted. The workings were bricked up in August, and digging did not restart for seven years. The tunnel was not finally completed until March 1843, when the great tunnelling shield was brought to the surface, its work done, and sold for scrap, much to the disgust of both Brunels, father and son.

After his adventures and brush with death in the tunnel, Brunel went first to Brighton to recuperate. In his journal he noted that he had met some 'pleasant company', but it seems that the excesses of this company did not have the desired effect, for he was then forced to take to his bed and rest fully, as the effects of the tunnel accident took their toll. His son Isambard records that, for much of the remainder of 1829, the engineer was 'without regular occupation' but occupied his time in scientific research and with correspondence with his friends Babbage and Faraday. By the summer of 1829 he had recovered enough to go away for his convalescence again, but this time the destination was the rather more sedate Clifton, a suburb of Bristol. It was not until the autumn,

however, that Brunel was to learn of a project which would effectively launch his engineering career — a proposed bridge over the Avon Gorge, close to his temporary home.

The Clifton Bridge had its origins in a bequest from a Bristol 'hooper' or wine merchant, William Vick, resident of the parish of All Saints, who, when he died in 1754, had left £1,000 to the Society of Merchant Venturers to accumulate interest until it reached £10,000. The Society was an ancient city guild dating back to 1551, occupied in promoting the interests of the City of Bristol. When the bequest total was finally reached, it was hoped that a stone bridge could be built over the Avon Gorge, linking Clifton with Leigh Woods on the Somerset side. When Vick died, both sides of the gorge were relatively underdeveloped but, by the beginning of the 19th century, Clifton was growing into a fashionable suburb, with rows of houses being built on the slopes of the gorge itself.

It soon became clear that the idea of constructing a stone bridge across the gorge was both impractical and excessively expensive. The gorge itself is over 700ft deep, and an added complication was the fact that at least 100ft clearance was needed for the safe passage of the many ships passing up and down the River Avon between Bristol and the sea. Some idea of the scale of the problem was given by the almost comical design submitted by the aptly-named William Bridges in 1759, which showed a six-storey construction, to include within it 20 houses, a corn exchange, chapel, library and museum, and a host of other features as well as a massive arch to allow ships through, the whole edifice supporting a 700ft roadway across the gorge.

By 1829 the fund had accumulated almost £8,000, and as a result a Bridge Committee was set up to consider the best

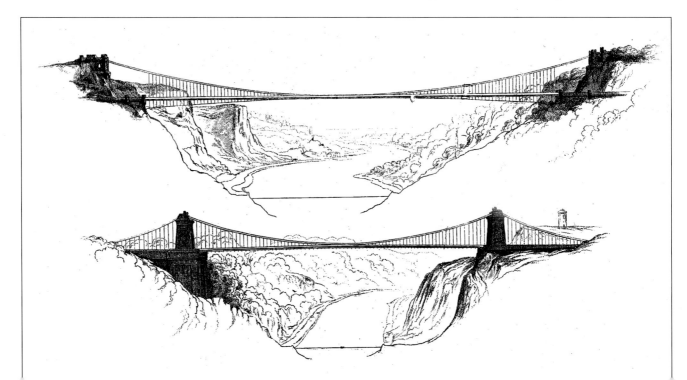

Far left: Two of Isambard's proposals for the Clifton Bridge; the most bold is the design in which the road plunges out of the rock face of the gorge itself on to the bridge. *Swindon Museum Service*

Top: One of the designs not accepted by the Bridge Committee. This proposal for a chain bridge was by William Armstrong. *Swindon Museum Service*

Above: A Victorian 'carte visite' of the Avon Gorge and Clifton Bridge. *Author's Collection*

Left: Detail from a sketch by Brunel for one of the bridge towers. *Swindon Museum Service*

way to implement Vick's bequest. There was by this time much public interest in the construction of a bridge over the gorge, no doubt fuelled by the success of Thomas Telford's bridges at Menai and Conway, both completed in 1826. The Committee was, however, somewhat taken aback to receive a quotation of around £90,000 to build a stone bridge, a sum clearly well out of its reach. It was decided therefore to abandon this scheme, and to adopt the idea of a suspension bridge, using the relatively new material, cast iron. On 1 October 1829 a competition to design a bridge was announced, with a prize of 100 guineas for the winner.

Brunel, after surveying the locality, initially submitted four drawings. Typically, at this early stage in his career, Brunel had no specialist knowledge of suspension bridge design, save that gained while working in his father's office in 1822, when Marc Isambard produced designs for bridges on the Ile de Bourbon (now Réunion Island) in the Indian Ocean. These bridges, although small in scale (the St Suzanne Bridge having a single span of 131ft 9in, the larger River du Mât two spans of the same size), gave Isambard valuable experience, particularly since they were designed to withstand hurricane-force winds.

Each of the Clifton proposals was to construct a bridge at a different crossing-point in the gorge, and each provided a different approach to the problem of spanning the river. Two designs were rather plain, and had little or no ornamentation. The first, with a span of 760ft, was not much liked by Brunel, since 'the location was not favourable to architectural effect', something which he felt the judges might take into consideration; however, it did have the advantage of being somewhat cheaper than the other options. The second design was situated some distance down the river, and again featured unornamented piers. Of more interest were the final two drawings, which proposed a far more daring idea. Both were situated in a position where the rocks of the gorge rose above the level of the bridge itself, and featured castellated towers supporting the structure, with the chains fixed directly to the rock. In one of the two designs, Brunel's favoured choice, the roadway on the Clifton side burrowed underground, emerging from a tunnel mouth directly on to the bridge itself.

By the end of November 1829 Brunel was one of only five architects to have been short-listed. A further 17 entries were rejected, many purely on financial or artistic grounds. 'Full of hope for his ultimate victory', Isambard had his hopes dashed by the announcement that the Committee, unable or unqualified to decide on the merits of the plans, had turned for advice to the respected engineer, Thomas Telford, who, as already mentioned, had designed the Menai and Conway bridges. A contemporary writer noted that Telford 'reported unfavourably on all the designs, considering that a greater span than 600 feet should not be attempted'. Much attention was focused at the time on damage to suspension bridges through wind causing excessive lateral movement. This, coupled to worries about the long-term stability of materials such as wrought iron used in suspension chains, led to the prevalent engineering opinion that risk increased with the span of the bridge. Even on his Menai Bridge, which had a span of less than 600ft, Telford had suffered from 'oscillation troubles', and he considered that the bridges of between 760ft and 1,160ft proposed by Brunel would be dangerous in a high wind.

Telford was then asked to produce a design of his own. This the Committee accepted, and it was his design which was submitted with a Bill for Parliamentary approval in January 1830. With the benefit of time, it is hard to find much positive to say about Telford's Gothic design; the faces of the bridge piers were to be covered in elaborate panelling, and the chains decorated with fretwork. As L. T. C. Rolt notes, it hardly seems possible that this 'monstrous aberration' could have been produced by the man who had so boldly spanned the Conway and Menai rivers. Having been produced, the design was then lithographed and sold, with 'thousands disposed of'. Brunel was, not unnaturally, disappointed at the news and wrote to the Bridge Committee, commenting critically on Telford's proposals and complaining at the Committee's timidity in choosing the design which reflected 'the state of the Arts in the present day'. The contest between Telford and Brunel could not have been more pronounced: Telford, 73 years old and in the twilight of his career, cautious and not a little jealous of his younger rival; Brunel, 24 years old, tackling his first major commission, and confident, arrogant even, of his own abilities and talent.

The supposed enthusiasm with which Telford's design was received was short-lived. Soon it seemed 'the captivating effect of the Gothic Belfries wore off'. Opposition was forthcoming from the Dean and Chapter of Bristol, who owned the Rownham Ferry, and claimed that the Bridge Trustees should pay them £200 per annum compensation, a claim vigorously countered by the Bridge Committee. Despite doubts as to the financial viability of the project, the Trustees were confident enough to go as far as purchasing land in 1829; Sir John Smyth of Ashton Court sold 4½ acres of land on the west side of the gorge for £1,107, and was roundly criticised locally as 'a despoiler of woods'. Parliamentary assent for the bridge was received in May 1830, but at this time there were still insufficient funds to construct Telford's elaborate scheme. Finally, in October, the Committee agreed to hold a second competition, this time excluding Telford altogether. Aiming to avoid controversy, it invited Davies Gilbert, a prominent scientist and former President of the Royal Society, to judge the competition, along with mathematician John Seaward. It is thought that the Committee chose Gilbert as he had suggested several important alterations to the Menai Bridge which had been adopted by a House of Commons Committee.

After the entries had been reduced to a short-list of four, Brunel once again found his plans rejected, this time in favour of designs by one W. Hawks. This time, however, he was not prepared to accept the verdict without a fight, and thus arranged a meeting with the judges and Trustees, to argue his case. Gilbert and Seaward had found fault with the design of the chains and their anchorage but, armed with drawings and sketches, Brunel persuaded them of both the technical and architectural merit of his scheme. Not only did he succeed in overturning the judges' decision, but he also managed to persuade them to accept his idea to decorate the towers in an Egyptian style; writing to his brother-in-law Benjamin Hawes shortly afterwards, he noted that this idea was 'quite extravagantly admired by all and unanimously adopted'. This unanimity, he argued, had been produced by 'fifteen men who were all quarrelling about the most ticklish subject — taste'. Not for the first time, Brunel had used his skills as an advocate and orator to persuade backers to adopt his scheme, a situation which will be apparent on more than one occasion later in this book. It seems that the Bridge Trustees were not averse to looking for ways to save money, even before Brunel had begun work on the project properly. Bridge historian A. E. Cotterell records that a Mr West, of the Clifton Observatory, thought that the adoption of wire cables might reduce the cost of the bridge, and travelled to France and Switzerland to see the Fribourg Bridge, which had been recently completed. On his return he attended a meeting with the Trustees, who also invited Brunel to attend. One can imagine that the engineer was less than happy at this invitation, and Cotterell records that he 'expressed dissent' and exclaimed that, if the Trustees wanted a cheaper bridge, he could design one with the same strength and durability, which would be capable of extension should funds be available. These plans, which would have cost £35,000 to implement, were initially approved by the Trustees, but this decision was later reversed, and Brunel's original proposals were adopted.

The initial optimism with which Brunel greeted his appointment as Engineer of the Clifton Bridge should be tempered by a more sanguine view, both of the finances of the project, and of the political and social upheavals which were about to sweep the country. From the beginning, the scheme was short of capital, a situation which improved little for many years and almost led to its complete abandonment. The design which Brunel eventually adopted involved positioning the Clifton bridge pier very close to the cliff edge, south of Observatory Hill. On the other side of the gorge, the Leigh Woods tower was to be built near an old Iron Age camp at Burwalls, but making the bridge level

required the construction of an enormous masonry abutment which jutted out into the Avon Gorge from its limestone slopes. By adopting this method, a span of 702ft was achieved, but the huge brick structure supporting the bridge must have added considerably to the cost of construction.

On 18 June 1831 work began to clear the site on the Clifton side of the Avon. Perhaps in the hope that this might persuade investors to contribute the £20,000 still needed to complete the scheme, a ceremony was held three days later. The success of the project was toasted with champagne and, after a band of Dragoon Guards had played the National Anthem, Lord Elton gave a speech, stating that Brunel would be seen as the 'man who raised that stupendous work, the ornament of Bristol and the Wonder of the Age'.

In his letter to Benjamin Hawes quoted earlier, Brunel had observed: 'if the confounded election does not come, I anticipate a pleasant job, for the expense seems no object providing it is made grand.' His optimism over the financial state of the project was exceeded only by his underestimation of the unrest in the country as a whole, and on 29 October 1831 Bristol saw what one city historian, John Latimer (writing in 1887), called 'the most disastrous outbreak of popular violence which has occurred in the present century'. Disturbances were sparked by the arrival in the city of the Recorder of Bristol, Sir Charles Wetherall — a man whom Latimer described as 'vehement and abusive' in his attacks on the Reform Bill. The learned judge escaped with his life as crowds grew increasingly hostile, and over the next three days the city burned, with the Mansion House, the Bishop's Palace and the New Gaol among the buildings looted and

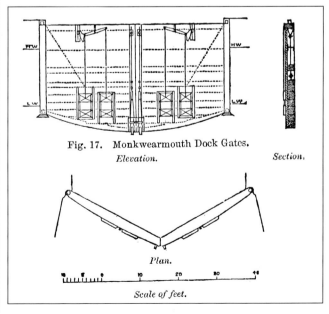

Fig. 17. Monkwearmouth Dock Gates.
Elevation.
Section.
Plan.
Scale of feet.

Above: Plan and elevation of the dock gates at Monkwearmouth, reproduced from the *Life of Isambard Kingdom Brunel,* written by his son. *Author's Collection*

burned. Brunel himself was a Special Constable, and rescued civic treasures from the wrecked Mansion House; however, his attitude towards the whole affair appears to have been rather ambivalent and, although he left a space to record his exploits, the pages in his journal are intriguingly left blank.

The riots stopped work on the bridge, and the collapse of confidence amongst potential investors meant that it was another five years before work could recommence. In 1832 Brunel had gloomily noted of the bridge: 'Nothing doing and no appearance of any likelihood of doing anything.' With little opportunity to continue with the project, Brunel was forced to look elsewhere for work and, on returning to London, was made aware of plans to build a complex of docks on the banks of the River Wear in Sunderland. Local businessmen were anxious to build more modern facilities to allow the rapid loading of coal which was being exported from nearby collieries. Travelling to the North East, Brunel produced designs for a 25-acre scheme, to be situated on the north side of the river at Monkwearmouth, which would include docks and warehousing. A rival scheme for docks on the south side of the river channel was designed by Francis Giles, and both came before Parliament shortly after. Brunel then enjoyed his first taste of the rough-and-tumble of committee cross-examination, a process with which he was to become more than familiar, when appointed as Engineer of the Great Western Railway. Although it was reported that Isambard acquitted himself well despite hostile questioning, in the event neither of the schemes was given Parliamentary assent. However, this was not the end of the story, and, after the defeat of the Monkwearmouth scheme, another company was set up to build a smaller dock on the site, to Brunel's design. Receiving its Royal Charter in 1834, the Monkwearmouth Dock Company built a much smaller six-acre dock which was completed two years later at a cost of £120,000.

The rejection of the original Monkwearmouth Dock scheme, ambitious though it was, proved to be something of a watershed in Brunel's career; up to this point it had been marked by a series of 'might-have-beens', with schemes such as the Clifton Bridge, Monkwearmouth Docks, a proposal for Woolwich Dockyard and his Gaz Engine all in abeyance and, worse still, not yielding any money. 'So many irons in the fire and none of them hot,' he wrote in his diary. After listing the various projects which he had begun, but had not completed, he noted one for which he still held out some hope — Bristol Docks. Brunel had been asked to make some suggestions for improvements to the docks to try and stem the steady decline of the port, which was falling behind its rivals such as Liverpool and Glasgow.

Although the original port had grown up around the tidal banks of the rivers Avon and Frome, in the early 19th century considerable money had been spent by the Dock Company to improve matters through the construction of the 'Floating Harbour', a new enclosed basin which maintained water at a constant high-tide level. The River Avon was given a new channel known as the 'New Cut', and entrance locks were built to give access from the river to the new facility. Despite this new harbour, which was nearly two miles long, there were still problems, the main being the fact that the harbour continued to silt up, and ships ran aground on mud and sediment washed in from both rivers. Brunel had been introduced to the directors of the Dock Company by Nicholas Roch, a friend and member of the Docks Committee. Asked to investigate the problem of silting, in August 1832 he produced a report which made a number of suggestions to improve matters. The first was to increase the flow of water through the Floating Harbour by raising the level of the dam at Netham, to force the maximum amount of river water through the docks. Sluices were also planned for Princes Street Bridge, as well as a culvert at the bottom end of the Floating Harbour. Arguing that these measures were no more than a short-term solution, he also proposed the use of a 'drag-boat' — essentially a flat-bottomed boat or barge, with a scraper on its keel, which could be regularly hauled across the harbour to scour the bottom. The mud could be dragged to a culvert where it could be washed out into the river.

Although the directors of the Dock Company received Brunel's report in September 1832, it was some considerable time before any action was taken. In February 1833 the Floating Harbour was drained and, after a tour of inspection, Brunel was able to meet the directors once again to make firm recommendations. After further delay, the directors finally approved the completion of only three of his recommendations, namely the sluices at Princes Street Bridge and at the far end of the Floating Harbour, and the construction and use of a 'proper drag-boat worked by steam'. Under the supervision of Brunel, the works were carried out in the following two years, with the drag-boat finally coming into use in the summer of 1834. Lack of capital prevented further work from being immediately carried out, and the level of the dam at Netham was not increased significantly despite complaints from the engineer. When asked to report on the state of the Harbour in 1842, Brunel was quick to point out to the directors that they had not heeded his original advice. 'I need hardly remind you that these preliminary measures were never adopted,' he wrote, concluding that the 'permanent interests of the Port were, I cannot but think, sacrificed to temporary convenience.' The directors, no doubt chastened by this stern lecture, softened somewhat and sanctioned further expenditure for various works including a larger drag-boat, but still refused to allow all his proposals to be undertaken. Brunel continued to be employed by the Dock Company, and mention of some of his other schemes, particularly in

Above: Two views of Brunel's South Entrance Lock to Bristol Docks, built between 1845 and 1849. Brunel had been asked by the Dock Company in 1844 to produce estimates for the new structure, but the company found it difficult to find the capital to cover the £22,000 it would cost to complete the work. Eventually finished in 1849, the lock was taken out of use in 1879 when a larger lock was opened to the north. The lock was 262ft long and 52ft wide, allowing larger vessels to enter the docks than previously. The swing bridge which crossed the lock is pictured in Chapter 5 of this book. *Author*

relation to the enlargement of the dock entrance, will be made in Chapter 4.

Brunel also found time to work on a number of dock schemes in the 1850s, most notably at Plymouth, not far from the Saltash Bridge, and at Briton Ferry in South Wales. Mention has already been made of his designs for the docks in Bristol, which were much delayed by lack of funds; at Plymouth, Brunel was employed to plan a new dock at Millbay, an inlet in Plymouth Sound. Work began in 1851, and was complete by 1856. Covering an area of 13 acres, the new dock basin had walls which were over 8ft thick at their base. The entrance to the harbour was protected by a pair of wrought-iron gates, each weighing over 75 tons. The gates were designed by Brunel with large air-chambers in them which meant that they were buoyant, an idea he had pioneered in his work at Bristol some years previously. Sluice valves in the gates also allowed the level of water in the dock to be regulated. As well as the larger 'wet dock', there was also a smaller 380ft-long graving dock, featuring a set of dock gates similar to those just described. In 1852 Brunel also designed a large 300ft pontoon, or floating pier, used to coal steamships at Millbay. Over 40ft wide, the pontoon could hold 4,000 tons of coal, and was connected to the shore by a twin-span iron bridge.

Another dock scheme in which Brunel was involved in the last years of his career was that at Briton Ferry in South Wales. A company had been set up to build a dock at the mouth of the River Neath at Baglan Bay as early as 1846, but little had been done until 1851, when Parliamentary permission was requested for work to begin. Matters were also helped with the passing of a Bill to build the South Wales Mineral Railway, which was to link the docks at Briton Ferry with coal mines at Glyncorrwg. Brunel was Engineer to this new line, as well as to the Vale of Neath Railway and the South Wales Railway, all of which were potential users of the docks. The docks themselves consisted of an outer tidal basin of 7½ acres and an inner harbour of 11 acres; connecting the two was a passage protected by a single dock gate of 56ft in length — at the time of construction, the largest single-leaf gate yet built. Like the gates at Bristol and Plymouth, it was of the buoyant type. Although work began on the project in 1853, construction only began in earnest in 1858, and Brunel did not live to see the scheme finished, Brereton overseeing its completion in 1861.

Above left: A view of the Millbay Dock at Plymouth, designed by Brunel and opened in 1856. For many years, Great Western Railway tenders ran between this dock and the large ocean liners which moored off Plymouth to disembark ocean mails and passengers. *Swindon Museum Service*

Left: Brunel's original pontoon built for the coaling of steamships at Plymouth, seen 103 years after its construction, in June 1955. *Swindon Museum Service*

Brunel's involvement with the Bristol Dock Company in the early 1830s did assist the development of his career, despite the fact that his plans were not fully implemented. As well as bringing him to the notice of commercial and business interests within Bristol which would eventually employ him to build the Great Western Railway, it also led him to meet Christopher Claxton RN, Quay Warden of the Port, a retired naval officer who was to become one of his closest friends and a key member of the committees responsible for the construction of his first two steamships, the *Great Western* and the *Great Britain*. Ironically it was the scheme to build the Great Western Railway which led to the revival of the Clifton Bridge, with Parliamentary assent for its Bill leading to a new air of optimism in Bristol.

Brunel had married Mary Horsley in July 1836, and had not long returned from his honeymoon when the first stone of the Leigh Woods abutment on the west side of the gorge was laid by the President of the British Association for the Advancement of Science, the Marquis of Northampton, on 27 August 1836. The ceremony was held at 7am to avoid disrupting traffic on the river, and the Clifton side was lined with people 'dressed in holiday attire', whilst the trees on the Leigh Woods side were decked with flags and bunting. One historian of the bridge noted that every spot in the locality 'upon which a human being could find a standing place was occupied'. Before the stone was laid, a copy of the Act of Parliament, various coins of the realm, and china, used by one Mr Ivatts of the Gloucester Hotel for a celebratory breakfast before the ceremony, were deposited behind the stone for posterity. Local tradition also has it that an attempt was made to perpetuate the memory of William Vick, through the selection of an inscription for the Leigh Woods pier. The phrase *Suspensa Vix Via Fit* is said to have been inscribed in what A. E. Cotterell called 'spurious Latin', and was translated as 'A suspended way made with difficulty'.

The feature which caused most interest was the suspension of a 1½in-thick iron bar across the gorge, to allow men and materials to be moved from one pier to the other. The bar itself was welded together in Leigh Woods and then hauled across to the Clifton side by capstan. Just when the bar was within reach, the capstan failed, and the bar plunged into the river. Although the bar was slightly bent, the process was repeated with success. Over 1,000ft in length, the bar was fixed into masonry on each side, and a basket attached. The basket ran to the centre by gravity, and was then hauled up each side by ropes. This rather perilous way of crossing the gorge proved popular with locals who, during periods when work was suspended on the bridge, paid to be hauled across. Writing in 1930, Clifton resident E. T. Lucas recalled his mother's memories of such crossings, which cost 5s per person. On one occasion a honeymoon couple, married at Failand on the Leigh Woods side of the bridge, were trapped

Above: The laying of the foundation stone for the Clifton Suspension Bridge on 21 June 1831. In the background can be faintly seen the steel bar suspended across the gorge, with the basket used to transport those brave or foolhardy enough to risk the journey about halfway across. *City of Bristol Museum & Art Gallery*

Below: Both towers are almost complete although still shrouded in scaffolding. With their completion, little else was done, and the bridge trustees were forced to sell off the suspension chains, there being little prospect of the bridge being completed. *City of Bristol Museum & Art Gallery*

for over an hour when the tow-rope snapped. Lucas reported that: 'Fortunately they had been supplied with a packet of sandwiches which came in very useful.'

By 1840 the Leigh Woods abutment had been completed, but not without further problems. In 1837 the main contractors working on the project were made bankrupt, and the Bridge Trustees were forced to run the project themselves, employing a combination of miners, quarrymen and other workmen to finish the work. In 1843 it seemed that the bridge was almost complete; the supporting towers were finished and awaiting chains, and much of the preparation, including approach roads, was complete. The financial problems which had beset the project since its inception could not, however, be forgotten, and in February 1843 it was revealed that the capital of £45,000 had been spent and a further £30,000 would be needed to finish the bridge.

All efforts to raise further funds were unsuccessful, and the very project which had proved the spur to investment in 1836, the Great Western Railway, now seemed to be draining away potential investors, who were attracted by the greater returns it appeared to offer. Even worse, there were still creditors to be paid, and in 1853 the Trustees were forced to sell off the chains and other equipment. The former were reused in another Brunel design, the Cornwall Railway's Royal Albert Bridge over the Tamar, at Saltash. In July 1853 work was completely abandoned, and for some years all that remained of Brunel's bridge were the rather forlorn bridge piers which stood overlooking the gorge unfinished, and unloved by many in the city. There had not been enough money to enable the Egyptian theme proposed by Brunel to be implemented; it was originally intended that the towers would be cased with cast-iron panels, illustrating the construction of the bridge from the quarrying of the iron ore to final completion. Two sphinxes on top of each tower were to have completed the design.

There was an abortive attempt to restart work in 1857, when the Trustees received a proposal to build a cheaper version of the bridge, using wire ropes instead of chains, from E. W. Serrell, an American engineer who had designed the Queenstown & Lewiston Bridge across the Niagara Falls. An agreement was made with a Bristol banker, J. W. Miles, to complete the bridge. This, Serrell claimed, could be achieved within 15 months of a new Act of Parliament being passed, and would cost no more than £17,000. Brunel was lukewarm about the scheme, and it did not proceed, Serrell being awarded £90 by the Trustees as compensation. Isambard's doubts were eventually vindicated, for in 1862 Serrell's bridge over the Niagara Falls blew down in a storm! It was not until 1860, a year after Brunel's death, that the two 'Lonely Giants' on each side of the gorge were finally rescued from the threat of demolition. The piers, known as 'Monuments of Failure', were to be part of a scheme to complete the project by a new company, set up by members of the Institute of Civil Engineers who wished to complete the bridge as a monument to the great engineer.

At a meeting held in May 1860 between the Trustees and the provisional committee set up to launch the new company, Captain Mark Huish announced that the latter had been set up to complete 'a monument to their late friend and colleague Isambard Kingdom Brunel, and at the same time removing a slur upon the Engineering talent of the Country'. With capital of £35,000, the company then reapplied to Parliament for assent to finish the bridge. Two great engineers of the time, W. H. Barlow and John Hawkshaw, were chosen for the completion of the Clifton project, and they restarted work with the towers largely complete, and some tunnels excavated for the anchorage of the suspension chains.

Matters were also helped by the fact that the suspension chains from Hungerford Bridge in London (designed by Brunel in 1842) became available when this was replaced by Charing Cross Railway Bridge. Hawkshaw was the Engineer for the latter, and no doubt arranged for the chains to be sold to the Clifton scheme for the very low sum of £5,000. The Hungerford Bridge had consisted of a central span of 676ft and side spans of 345ft; the chains were similar in design to those proposed by Brunel for Clifton, and this bittersweet coincidence helped smooth the passage of work on the revived project.

A new company prospectus was issued in May 1860 and sent to original subscribers to the bridge project. The Secretary, Christopher Claxton, wrote that, in completing of Brunel's bridge, the promoters would 'erect a monument to his fame of the noblest kind'. Realising, however, that mere sentiment might not be enough to persuade Bristolians to part with money for a project which had already failed once before, the letter concluded with the boast that 'from the advantageous character of the financial arrangements entered into it is confidently expected that the project will prove remunerative'. It is clear from contemporary accounts in the local press that the completion of the bridge was also seen as a boost to the reputation of the city. The *Bristol Mirror* reported in May 1860 that: 'We in Bristol have our own credit at stake. The bare piers at Clifton and Leigh have long been a reproach to us and the subject of comment by thousands of visitors.'

In November 1860 the Clifton Suspension Bridge Company issued a report which gave details of how the bridge would be finally completed. A further 500 tons of chain were necessary, as Barlow and Hawkshaw decided to use three chains rather than the two proposed in the original designs. It was also decided to increase the width of the roadway from 24ft to 30ft, which was estimated to cost an additional £5,000. The report also noted that arrangements had been made with Sir Greville Smyth, a 'large landed

proprietor' on the Somerset side of the gorge, who would purchase £2,500 worth of shares and also give a £2,500 'donation'. For this gift, Sir Greville was given the privilege of exemption from tolls for a period of 30 years from the opening of the bridge. In June 1861 a new Act of Parliament was passed, with work starting exactly a year later. This time sufficient capital was raised, allowing the employment of Cochrane & Co as main contractors.

A symbolic and practical goal was achieved in June 1863 when a temporary bridge was strung between the two towers, ending their gaunt isolation. Writing in a guide to the city some years later, James Baker described the scene on the day of the installation of this structure, remarking that one of the workmen was seen to 'jump on this bare plank and shout "hurrah!" without showing any fear' — no mean feat, considering the river lay 250ft below! Each set of chains was independent of the others and consisted of flattened links, with the ends enlarged, to accommodate bolt-holes. Arranged side-by-side, each link was interlaced with the next. The assembly of the chains began from the anchorages at either end of the bridge, with around 100 links being joined together in a day. Work began on this in August 1863, when both the anchorage and the staging and supports to carry the chains had been completed. By May 1864 the chains were complete.

Construction continued steadily until July 1864, when the last part of the roadway was fixed in the centre of the bridge; the floor of the roadway was built from 5in-thick Baltic timber, over which a further layer of 2in timber was laid transversely. Before the bridge could be opened it was tested with a load of 500 tons of stone, spread evenly across the road and pavements. The deflection measured was 7in in the centre and, when the load was removed, it returned to within ⅟₁₆in of its original position. Contemporary sources calculated that a load of 28,000 tons would be required to cause the bridge to fail, a situation which Baker noted meant that 'no one passing over it need have a fear of receiving a bath in the Avon'.

The opening of the bridge on 8 December 1864 was without doubt one of the most impressive public events to have been seen in the city for many years, the local newspapers estimating that around 150,000 people lined the streets to watch the proceedings. Heavy rain had fallen the night before, the *Bristol Daily Post* reporting that 'men's hearts sank within them' as they realised preparations for the opening had 'labours in vain'. The weather did, however, improve, and at 10.00 in the morning a huge procession made its way from the city centre, accompanied by the army and at least 16 bands. There was some

disappointment locally that no member of the Royal Family had attended the opening, the *Bristol Daily Post* noting that 'the Prince of Wales or Queen Victoria herself would have lost no dignity by being present'.

The bridge and the surrounding area were heavily decorated with evergreens, artificial flowers and flags, and a number of triumphal arches were erected on the Clifton side. The ceremony commenced at noon with the first crossing, accompanied by cannon-fire from the Leigh Woods side of the bridge. After the usual speeches, the invited dignitaries retreated to the nearby Victoria Rooms to enjoy a ceremonial banquet, while the local population remained to see the bridge illuminated in the winter darkness. This by all accounts did not quite produce the desired effect. It was reported that electric lights were positioned on the top of each bridge pier, with magnesium lamps at intervals along the roadway. When all worked together, the effect was striking, but the wind and the unreliable nature of some of the lights gave, it was noted, 'a dim appearance and caused great disappointment'.

The bridge as finally completed stands not only as a monument to the flair and determination of Brunel, but also as a mark of the respect accorded him by his contemporaries. Although the completed structure was rather less grand than perhaps he might have wished, the austere but imposing bridge piers still carry an echo of what Brunel called his 'Egyptian thing', and the soundness of his original designs has meant that the Clifton Suspension Bridge will continue for many years to be a landmark in the city which helped launch his career. Brunel's son wrote that his father 'never forgot the debt he owed to Bristol, and to the friends who helped him there'. Had his designs not been accepted by the Bridge Committee, it is likely that he would have found fame sooner or later, but the Clifton Bridge gave him the opportunity to make his name, and to go on to yet-more-ambitious projects.

CLIFTON SUSPENSION BRIDGE. by.

Above: The opening of the Clifton Suspension Bridge in 1864 as depicted by the *Illustrated London News.*

Left: The elegant lines of the completed bridge are shown in this Edwardian postcard. *Author's Collection*

Right: This painting of an early Great Western locomotive between Bristol and Bath clearly shows the nature of Brunel's track. *National Railway Museum*

Chapter Two

The Birth of the
Great Western Railway

Writing in June 1860, George Measom, author of *The Official Illustrated Guide to the Great Western Railway*, noted that the line could be ranked 'among the hundred or more marvels that Great Britain has produced'. 'The genius of Brunel,' he continued, 'a name imperishably stamped upon the scroll of fame, stands associated with the vast undertaking'. Written a year after the great engineer's death, the author's hyperbole can be forgiven, but nevertheless it is in the design and construction of the Great Western Railway that Brunel achieved the fame and success which had hitherto eluded him. By the end of 1835, with Parliamentary assent for the railway granted and construction work underway, he was able to note in his diary that he was 'the engineer to the finest work in England'. Although he had grounds for optimism, he was quick to acknowledge that it had not been an easy year; the birth of the railway was not painless, and Brunel had played a key role in steering it through its early years.

The promotion of a new railway from Bristol to London owed much to the efforts of the Merchant Venturers, who only a few years before had played such an important part in the Clifton Suspension Bridge scheme. There was good commercial sense in linking the city with London; Bristol was seen as the most important port in the West Country and, if it were to sustain the prosperity which had been foundered on tobacco and the slave trade in the 18th century, it needed good communications with the capital.

Although turnpike roads had improved much in the 50 or so years prior to the construction of the Great Western, the journey between Bristol and London could take over 16 long hours, with changesof horses at regular intervals en route. More important to the Merchant Venturers, however, was the high cost of goods haulage by road. The opening in 1810 of the Kennet & Avon Canal created a waterway which linked the two great cities; however, shortages of water, and the delays caused in winter when the canal was frozen, meant that matters were less than satisfactory. The construction of the canal had been much delayed by difficulties in raising enough capital to complete the scheme, particularly since extensive cuttings, embankments and bridges were required at the west end of the canal, a situation later mirrored to a certain extent in the Great Western Railway project.

Numerous abortive railway schemes had been proposed to link Bristol and London in the years before 1830, the earliest, dating from 1800, being a horse-drawn line proposed by Dr James Anderson. In 1824 John McAdam, the famous road-builder and surveyor of the Bristol Turnpike Trust, promoted the London & Bristol Rail Road Company, which was to run from Bristol through Mangotsfield, Wootton Bassett, Wantage and Wallingford, with its London terminus being many miles from the centre

Below: The city of Bristol, as depicted by J. C. Bourne in 1846.
Swindon Museum Service

of the city, in Brentford. Despite public meetings the scheme came to nothing, as did a similar venture mooted the following year by one Francis Fortune for the 'General Junction Railroad'. Other, more limited railways linking Bristol with Bath, and London with Reading, were also discussed, but, like the others, they came to nothing.

The opening of the Stockton & Darlington Railway in 1825, and the more general growth of railways in the North East of England, led the directors of the Kennet & Avon Canal to send their engineer north to investigate in 1828, and it was hoped that lessons learned could be incorporated in some speculative scheme. Although the benefits of rail over canal must have been more than apparent, it was the construction and opening of the Liverpool & Manchester Railway some two years later which had by far the greatest impact nationally. The success of Britain's first inter-city railway 'changed the opinions of scientific men', wrote Dionysius Lardner in 1835; 'it came upon the scientific world like a miracle'. Stephenson's railway showed that steam locomotives could be used to transport goods and passengers over long distances, and opened the way for other lines between major cities, most notably London and Birmingham, and, once more, London and Bristol.

The upsurge of interest in railways resulted in the promotion of yet another scheme, this time with rather more vigour than before. The Bristol & London Railroad, which issued a prospectus in May 1832, was to run between the two cities on a southerly route, passing through Bath, Bradford-on-Avon, Trowbridge, Newbury and Southall, with a London station 'within three or four hundred yards of Edgware Road, Oxford Street and the Paddington & City Turnpike'. Rather confusingly, a month later the promoters, engineers William Brunton and Henry Price, issued yet another prospectus, this time for a London & Bristol Railway. The £2,500,000 cost of the line may have been one reason for its failure to raise enough capital but, in all events, its demise was not the end of the story. In the autumn of the same year, four Bristol businessmen, Thomas Guppy, John Harford, George Jones and William Harford, met to consider once again the question of a railway to London. Having agreed on the importance of the idea, they set about persuading those with influence and power that their idea was a good one. In this they succeeded, and on 21 January 1833 a meeting between the Merchant Venturers, Bristol Corporation, the Bristol Dock Company, the Chamber of Commerce and the Bristol & Gloucestershire Rail Road Company was called to 'take into consideration the expediency of promoting the formation of a Rail Road from Bristol to London'.

With such serious support from five of the most influential groups in the city, any such scheme had a far better chance of success than many of the purely speculative projects described previously. These five bodies agreed jointly to fund a survey of the route, and deputed Nicholas Roch, a member of the Bristol Docks Committee, to select an engineer for this purpose. Roch was a friend of Brunel's, having worked closely with him on the dock improvements described in Chapter 1. He told Brunel of the project on 21 February 1833, exactly one month after the first meeting of the Bristol committee. Brunel was, however, in competition with a number of rivals, all of whom had already played an important part in railway development in Bristol. Perhaps the best known were Brunton and Price, promoters of the 1832 Bristol & London Rail Road scheme. The other was W. H. Townsend, a local surveyor and engineer who had been responsible for the survey and design of the Bristol & Gloucestershire Railway. This short line was in fact the first railway to enter the city of Bristol, opening five years before the Great Western, in August 1835. The railway was a horse-drawn tramway with cast-iron rails on stone blocks, which linked the coal mines at Coalpit Heath with the River Avon at Cuckold's Pill. Important though the railway was locally, it was obvious that Townsend had only limited experience of railway construction, and probably lacked the ability to survey a major new railway linking two of the country's biggest cities.

The railway committee was determined that the method of selection should involve each of the rival engineers surveying a route, with the lowest estimate winning the job. In characteristic fashion, Brunel told Roch that this method was completely unacceptable — he would survey a route that was the best, not the cheapest. Committing his thoughts to paper for the benefit of the committee, he wrote: 'You are holding out a premium to the man who will make you the most flattering promises.' Not for the first time, Brunel boldly gambled with his reputation, and it should be remembered that, whilst undoubtedly a talented engineer, Brunel had little practical experience of railways himself at this point, and was quite probably hoping that the good impression he had made in Bristol with his Clifton Bridge scheme would sway the promoters. While the committee considered his proposition, he travelled back to London to attend the Annual Meeting of the Thames Tunnel Company. He returned to Bristol two days later, on 6 March, to discover that the committee had confirmed his appointment, with Townsend as his assistant. On dining with Roch the next day, Brunel learned that his appointment had been approved by one vote — rather too close a call even for him, who noted that he should 'be more active next time'.

There was no time to waste, since the committee had decreed that a preliminary survey should be completed within a month. Much of the work was done on horseback, and Brunel's diaries record his travels far and wide across the proposed route of the line. As Engineer of the Bristol & Gloucestershire Railway, Townsend appears to have suggested that the new railway should mirror the route of the tramway as far as Bath. Brunel dismissed this rather circuitous plan, preferring to follow instead the valley of the

Top: Temple Back in Bristol, where the first meeting to form the Great Western was held. *Swindon Museum Service*

Above: The formal invitation to the public meeting to discuss the railway project in January 1833. *Swindon Museum Service*

River Avon. Further east, he surveyed two routes, one running through the Vale of Pewsey and the Kennet Valley to Reading, the other through the Vale of White Horse to Swindon and Wootton Bassett. Brunel recommended the latter, the more northerly of the two, and estimated the cost of construction to be £2,800,000.

Shortly after the completion of this survey, the committee formally launched the project at a public meeting, held on 30 July 1833 at the Bristol Guildhall.

Above: The Guildhall in Bristol, where the first public meeting of the railway was held in 1833. *Swindon Museum Service*

Below: The scene at the first public meeting of what became known as the Great Western Railway, as recreated for the GWR centenary film in 1935. *Swindon Museum Service*

After some discussion, it was decided that a company should be formed to construct the as-yet unnamed railway. The meeting further resolved that a 'General Board of Management' should be created, consisting of directors from Bristol and London. The official historian of the Great Western Railway, McDermot, recorded that no fewer than 30 Bristol directors and deputies were elected; however, when the first prospectus of the company was issued at the end of August, this rather unwieldy number had been reduced to 12 directors from each city. Brunel was called to speak at the meeting and, interestingly, for someone so forthright, it was a task he did not relish. 'Got through it tolerably,' he noted; 'I hate public meetings — it's playing with a tiger, and all you can hope is that you don't get scratched.'

To complete the process, a London committee was formed, largely through the influence of George Henry Gibbs, head of Antony Gibbs & Sons, and cousin of George Gibbs, a Merchant Venturer and a member of the Bristol committee. It was in the offices of Gibbs & Sons, at Lime Street in the City of London, that the first joint meeting of the London and Bristol committees was held, on 22 August 1833. Shortly afterwards the company's first prospectus was issued, and with it came the first use of the

Above: This 1889 view of the carpenters' shop at Swindon Works reveals that Brunel's 'Flying Hearse' was still in existence then, and can be seen on the right-hand side of the photograph. *Swindon Museum Service*

Below: The coat of arms of the early Great Western company. *Swindon Museum Service*

name 'Great Western Railway', rather than the more clumsy Bristol & London Railroad used hitherto. The capital needed for the new railway was estimated at £3,000,000, to be financed in shares of £100 each. This formidable amount proved difficult to raise, and by October 1833 less than a quarter of this sum had been reached, despite the efforts of the directors and the Secretary of the London committee Charles Saunders, who had travelled extensively, canvassing support from members of both houses of Parliament. Brunel had first met Saunders on 27 August, pronouncing him to be 'an agreeable man'. Saunders was 36 years old, and was an able and well-travelled businessman who was eventually to play a key role in the development of the Great Western as its Company Secretary.

Much attention had been paid to the administrative and legal framework necessary for the promotion of the new railway; Brunel no doubt found much of this frustrating, and his patience was rewarded on 7 September, when it was decided that the detailed survey of the line should start in earnest. During the autumn, Brunel and a number of newly-appointed assistants travelled the length of the route. Although the additional staff employed were of assistance, much of the responsibility for the completion of the survey rested with Brunel, and his diaries record something of the enormity of the task. Although he had ordered a carriage which would double as a mobile office, a black britzska nicknamed 'the flying hearse', Brunel spent most of his days on horseback, checking on the survey and visiting landowners whose property was likely to be crossed by the new railway. Even when he paused at night to lodge at an inn, his work was not completed, with reports and letters to be written before he could sleep. 'It is harder work than I like,' he wrote to his assistant Hammond; 'I am rarely under twenty hours a day at it.'

Since no railway Bill could proceed through Parliament without half its capital having been subscribed, the directors decided that drastic measures were required. On 23 October they announced that they intended to apply to Parliament to build a railway from London to Reading (with a branch to Windsor), and from Bristol to Bath, 'thereby rendering the completion of the whole line more certain', with a further application to complete the missing link to be made the following year. Brunel was instructed to abandon work on the Reading-Bath section, but the plans for the two proposed ends of the new railway, completed and put before Parliament in November 1833, showed a railway which was very similar to that completed in 1841, except for differences at the London end, where it was to run on a four-mile-long viaduct from Vauxhall Bridge, through Pimlico, Brompton, Hammersmith and South Acton. After a small tunnel south of Ealing, the railway threaded its way westwards through West Drayton, Slough, Maidenhead and Twyford. Instead of the cutting eventually built through Sonning Hill, Brunel proposed a tunnel, before the line terminated at Reading.

The second reading of the Great Western Railway Bill in March 1834 gave Brunel another new challenge. After a House of Commons debate lasting several hours, the Bill was passed by 182 votes to 92, and then moved to the Committee stage. Chaired by Lord Granville Somerset, the Committee first met on 16 April, and then sat for a staggering 57 days. L. T. C. Rolt remarks that this epic Parliamentary struggle was remarkable for two reasons: the amount of nonsense spoken by opposition witnesses, and the skill and patience of Brunel under cross-examination. Francis, in his *History of the English Railway* published in 1850, remarked that the opposition to other lines extended to the Great Western, and that all manner of objections were raised. Apart from the usual one that the line was the speculation of 'engineers, attorneys and capitalists', it was alleged that 'people would be smothered in tunnels...engines would be upset, necks would be broken'. Other fantastic claims included one that the water supply for Windsor Castle would be destroyed, and another from a farmer who objected because he feared his

Above: Charles Saunders, first Secretary of the Great Western Railway and one of Brunel's staunchest allies. *Swindon Museum Service*

cows would be killed passing under a railway bridge.

Brunel, as the line's engineer, was the chief witness for matters of a technical nature, and he endured an 11-day barrage of questions, some straightforward, others absurd. Although there was a good deal of support for the railway, particularly in terms of its advantages for both passenger and goods traffic, it faced formidable opposition from a number of influential opponents. Early objections from landowners in the Vauxhall Bridge area meant that the last two miles of the line were not proceeded with, the railway instead terminating at the Old Brompton Road, a decision not greeted with any enthusiasm by residents there.

More concerted opposition came from landowners in the Windsor area, for a variety of reasons. The residents of the Royal Borough were against the railway because it did not run close enough to the town; the Provost of Eton College had a quite different view, arguing that no public good could come from such an undertaking as the Great Western. The railway would be 'injurious to the discipline of the school, and dangerous to the morals of the pupils' and, Francis added, it was noted that 'anyone who knew the nature of Eton Boys, would know that they could not be kept from the railway'. A more serious threat came from the

Great Western Railway,

BETWEEN
BRISTOL AND LONDON.

Deposit £5 per Share.

UNDER THE MANAGEMENT OF A
BOARD OF DIRECTORS,
CONSISTING OF THE

LONDON COMMITTEE.		BRISTOL COMMITTEE.	
JOHN BETTINGTON, Esq.	ROBERT HOPKINS, Jun. Esq.	ROBERT BRIGHT, Esq.	WM. SINGER JACQUES, Esq.
HENRY CAYLEY, Esq.	EDW. WHELER MILLS, Esq.	JOHN CAVE, Esq.	GEORGE JONES, Esq.
RALPH FENWICK, Esq.	BENJAMIN SHAW, Esq.	CHAS. BOWLES FRIPP, Esq.	JAMES LEAN, Esq.
GEORGE HENRY GIBBS, Esq.	HENRY SIMONDS, Esq.	GEORGE GIBBS, Esq.	PETER MAZE, Esq.
ROBERT FRED. GOWER, Esq.	WILLIAM UNWIN SIMS, Esq.	THOS. RICH. GUPPY, Esq.	NICHOLAS ROCH, Esq.
RIVERSDALE W. GRENFELL, Esq.	GEORGE WILDES, Esq.	JOHN HARFORD, Esq.	JOHN VINING, Esq.

C. A. SAUNDERS, Esq. Secretary.

Office, No. 17, Cornhill.

W. TOTHILL, Esq. Secretary.

Railway Office, Bristol.

Bankers { LONDON :— Messrs. GLYN, HALLIFAX, MILLS, & Co.
BRISTOL :— { Messrs. MILES, HARFORD & Co.
Messrs. ELTON, BAILLIE, AMES & Co.
Messrs. STUCKEY & Co.

Solicitors { LONDON :— Messrs. SWAIN, STEVENS & Co.
BRISTOL :— Messrs. OSBORNES & WARD.

Engineer, J. K. BRUNEL, Esq.

Applications for Shares to be addressed to the Secretary in London or Bristol, from whom the Prospectus may be obtained.

Subscribers will not be answerable beyond the amount of their respective Shares.

The establishment of a Railway for the important purpose of connecting the Western Districts of this Country with the Metropolis has been resolved upon, as the result of a minute investigation, conducted by the Municipal Corporation of Bristol in concert with the other Public Bodies of that City.

The general advantage of Railways is no longer to be considered a speculative theory. They have invariably conferred an additional value upon the property contiguous to them ; and this fact alone establishes their claim to be considered as National undertakings, contributing to the permanent Wealth of the Country.

The Great Western Railway is recommended also by peculiar local advantages.

The actual amount of travelling on the Line of the projected Railway is even greater than is commonly supposed : it exceeds, in fact, that on any other road to an equal distance from London : it forms the communication of the metropolis and its vicinity Westward, with Windsor, Maidenhead, Reading, Oxford, Cheltenham, Gloucester, Bath and Bristol. Daily communication to a very large extent, takes place by Coaches from the latter places to the West and South-west of England, including the clothing districts of Wilts, Somerset and Gloucester, Wells, Bridgewater, Taunton, Exeter, Plymouth, Devonport, Falmouth, &c.: and by Steam-boats, with the Ports of the Bristol Channel, South Wales and Ireland. On considering these various and extensive sources of Revenue to the Railway, the amount, subsequently stated, will excite no surprise.

The additional facility of intercourse afforded by the establishment of a Railway, will infallibly multiply the present large number of travellers. That mode of communication will also afford an improved conveyance of goods: it will encourage manufactures in the vicinity of the coal fields which surround Bristol, and in both these ways promote the commerce of that Port: it will diffuse the advantages of the vicinity of towns over the whole tract of country intersected by it: it will improve the supply of provisions to the Metropolis, as well as to those towns, and extend the market for agricultural produce; it will also give considerable employment to the laboring class both during its construction and by its subsequent effects; and will enhance the value of property in its neighbourhood.

Above: Eton College c1852. *Author's Collection*

Left: Title page of the Great Western Prospectus, 1834. *Swindon Museum Service*

Below: The route of the original Great Western Railway, from the 1834 Prospectus. *Swindon Museum Service*

rival London & Southampton Railway, which was seeking Parliamentary assent for a bill in the same session, a threat which was to return the following year.

Having heard all the evidence in this marathon session, the Committee finally approved the Great Western Bill, returning it to the House of Commons. The scheme was, however, rejected by the House of Lords by 47 votes to 30, on 25 July 1834. 'Neither "Great" nor "Western"' was how one of the opposing counsels described the original 1834 proposals. 'To halt a western railway (sic) at Reading was absurd', Francis declared, and the company must 'have a complete western railway or none'. Although some saw the defeat as the end of the scheme, the company directors were undeterred, and the victory celebrated at a public meeting at Salthill by opponents like the Marquis of Chandos could only be short-lived. During those 57 long days, many of the arguments had been won, and public opinion was now firmly behind the Great Western.

A new prospectus was issued in September 1834, this time for the complete railway. The route of the line was confirmed as running through Slough, Maidenhead, Reading and Wantage, continuing through Swindon, Wootton Bassett, Chippenham, Bath, and on to Bristol. Brunel and the directors felt it necessary to state firmly why this route had been chosen in preference to a more southerly alignment through Hungerford, Devizes and Bradford-on-

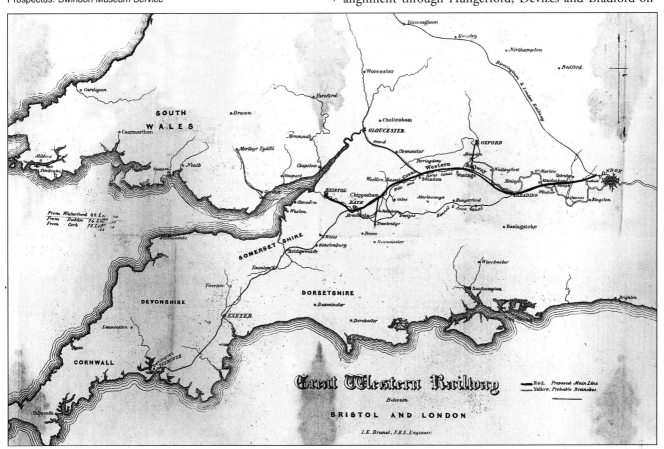

Avon. The prospectus noted that the terrain might cause 'difficulties and expense', although it is hard to imagine that engineering difficulties in the area could have been any worse than those encountered between Swindon and Chippenham on the chosen route. More important was the access which the more northerly line gave to Oxford, Cheltenham, Gloucester, Stroud and the Gloucestershire wool trade. Beyond Gloucester, of course, was South Wales, and the Great Western directors were not blind to the potential revenue from this thriving industrial area.

The revised capital for the 116-mile line was now exactly £2,500,000 — a reduction of over £300,000 due in the most part to a revised route into London. No terminus in the capital was indicated in the prospectus since at the time of its issue, negotiations were in hand to share the facilities of the London & Birmingham Railway at Euston. When this arrangement was confirmed in November, yet another revised prospectus was issued. Meanwhile the company busied itself in raising the additional capital necessary for a Bill to be presented to Parliament. Public meetings were held in towns and cities throughout the West Country, and both Brunel and Charles Saunders travelled extensively, canvassing support wherever they could. Saunders in particular worked tirelessly in raising the required capital, and his efforts paid off when, at the end of February 1835, he was able to announce that capital of £2,000,000 had been raised.

The new Bill for the Great Western Railway was duly submitted to Parliament, and those hoping for an easier passage this time were initially heartened by a decision made early in the Committee proceedings. Charles Russell, Committee Chairman and MP for Reading, announced that, since the advantages and merits of the railway had generally been agreed in the 1834 Committee report, this need not be debated again, and that any discussion should centre around the route proposed. This decision did not prevent a further 40 days in Committee, with Brunel again the chief witness for a large proportion of the proceedings. Although the objections of Eton College still prevailed, there was more serious opposition from the London & Southampton Railway, which had proposed a rival scheme, the Basing & Bath Railway, which would have followed the more southerly route described earlier.

Having already lost the argument with regard to the necessity of the Great Western itself, Brunel's opponents were forced to argue the merits of their own plans. Perhaps to deflect attention from this, much wrath was vented on the engineer's plan to build a tunnel through the hill at Box. This structure was deemed 'monstrous and extraordinary', witnesses describing how the construction of the tunnel would inevitably lead to 'the destruction of human life' and that the noise of two trains passing in the tunnel would

Above: Handbill for the rival Basing, Bath & Bristol scheme. *Swindon Museum Service*

'shake the nerves of this assembly'. Perhaps the most outlandish claims came from Dionysius Lardner, an Irish scientific writer who had edited the 139-volume *Cabinet Cyclopædia* in 1829, but whose often unconventional and ill-founded ideas made him a figure of ridicule in the scientific community as a whole. Lardner appears to have enjoyed criticising some of the most important figures of the day, and after making disparaging comments about the inventor Samuel Hall, he was denounced as 'ignorant and impudent' at a meeting of the British Association. Lardner produced calculations to show that a train running without brakes down the 1 in 100 gradient of Box tunnel would increase speed, reaching the west portal travelling at 120mph! Brunel was able to point out that his sparring partner had neglected to include friction and air resistance in his calculations, reducing the estimated speed to 56mph. This was not the last Brunel was to hear of Lardner, who also had strong views about steamship technology, and his intervention during the promotion of the Great Western Steamship Company in Bristol, described in Chapter 4 of this book, was to cost the company and Brunel dear, both in terms of actual investment and in shareholder confidence.

At the end of August 1835 the Committee finally finished its deliberations, and returned the Bill to the Lords with a number of concessions aimed at mollifying the authorities of Eton College, most notably that no station was to be built within three miles of the school, and that a sturdy fence should be erected alongside the line in the vicinity. With these additions, the Great Western Railway Bill eventually received Royal Assent on 31 August 1835. After a tumultuous struggle, Brunel's railway was finally a reality. The process had not been easy, neither was it cheap; in a report given to the first general meeting of the proprietors of the company on 29 October 1835, it was noted that, out of a total of £88,289 spent by the company to that date, over £54,000 had been used to fund the Parliamentary fight. The directors were satisfied, however, that 'the expense thus incurred may fairly be considered by no means disproportionate to the object attained'.

Less than a month later, work had begun on the new project, albeit on a limited scale. Brunel wrote to his old partner Townsend, asking him to get undergrowth cut down at Brislington near Bristol, to enable the alignment of the line to be set out. By November the first contract had been placed, for the great viaduct at Wharncliffe, and over the following year construction began at both ends of the line. With this underway, something of Brunel's vision for his new railway began to emerge. It soon became obvious that this would be no carbon-copy of what had gone before on other lines such as the Liverpool & Manchester, for, although railway engineering was still in its infancy Brunel's vision was one of a complete railway system, which would include innovations in architecture, trackwork, motive power and, importantly, track gauge.

When questioned later, Brunel could not recall when he made the momentous decision to adopt a track gauge of 7ft rather than the 4ft 8½in already in common use. 'I think the impression grew upon me gradually,' he told the Gauge Commission in 1845. It seems likely, however, that his thoughts crystallised sometime during the hiatus between the reading of the 1834 and 1835 Parliamentary Bills; Rolt records that the first Great Western Bill (of which no copy appears to have survived) contained a clause restricting the railway to 4ft 8½in. However, when the revised Bill came before Parliament, any such clause was absent, Brunel having persuaded Lord Shaftesbury, Chairman of Committees in the Lords, to drop it, using, rather ironically, the precedent of the London & Southampton Railway Bill, which for some reason had no mention of track gauge in it. The first news of this momentous decision was received by the directors in a special report submitted to them on 15 September 1835, in which Brunel recommended with regard to the width of the rails 'a deviation from the dimensions adopted in the railways hitherto constructed'. It

is debatable how much the directors would have gleaned from the highly technical report, dealing as it did with resistance, friction and wheel size.

Far from extolling the virtues of the broad-gauge, Brunel's report only cited its advantage in terms of reducing the centre of gravity of rolling stock, by mounting coach and wagon bodies between the wheels rather than above them. In any event this proved to be a rather dangerous and impractical method of construction, and few vehicles of this type were built or used. There was no mention in the report of the other advantages of the broad-gauge, namely larger and more powerful locomotives and roomier rolling stock. The report did end with a brief summary of the disadvantages of the system, which Brunel argued could require some increase in the size of earthworks, bridges and tunnels, slightly heavier carriages, and, most importantly, 'the inconvenience of effecting the junction with the London & Birmingham Railway'.

The whole question of the siting of the London terminus of the Great Western had already caused difficulties to the company, and the arrangement noted in the 1835 Bill to share a station at Euston with the London & Birmingham came to grief soon after the passing of the Act in August 1835. Much has been written regarding the abandonment of this scheme, with most commentators blaming Brunel's adoption of the broad-gauge as the main reason for the breakdown in negotiations between the two companies. It appears however that this argument may be rather simplistic and, although the adoption of the broad-gauge was a major factor in souring relations, it seems that the refusal of the London & Birmingham to allow the GWR to purchase (or lease on a long term) land and access to the junction was the major stumbling-block. In the report produced for the first half-yearly meeting in February 1836, the company noted that the London & Birmingham would only offer a five-year lease on land and buildings, a situation which would have allowed it to hold the Great Western to ransom at the end of five years, demanding 'higher pecuniary terms', as the report put it. Negotiations were abandoned, the company 'entertaining also a confident belief that an excellent and independent terminus can be secured without much difficulty'. A site for a new station was found at Paddington, and new Parliamentary powers were requested to allow this to take place, with a new line of around 4½ miles running from Acton to the new terminus.

The end of this particular saga was marked with some recrimination, particularly from the London & Birmingham, which as a result of the disagreement was left with rather more land than it really required at Euston and Camden Town. McDermot noted that Robert Stephenson, the line's Engineer, argued that, had the Great Western retained the standard-gauge, other legal matters might well have been resolved. For its part, the Great Western was no less forthright, Charles Saunders reporting that: 'The London & Birmingham were glad to get rid of us.' Not long after, Parliamentary assent was also granted for the Cheltenham & Great Western Union Railway, which would run north from Swindon, in the face of opposition from the London & Birmingham, which had promoted its own railway to Cheltenham via Tring and Oxford.

The question of gauge was debated and finally accepted by the GWR Board on 29 October 1835, but, while details of the decision were public knowledge, this was not reported officially to shareholders until 25 August 1836. Within the Board's description of work in progress on the new line, it was noted that 'to remedy several serious inconveniences experienced in existing railways, an increased width of rails has been recommended by your engineer, and after mature consideration has been determined upon by the Directors'. This seal of approval was by no means the end of the story, as we shall see later in the book. John Francis, writing some years after the whole gauge controversy, had a rather more cynical view of Great Western shareholders who, he argued, 'beheld titanic arches and vast tunnels, magnificent bridges and fine viaducts... their increased expenses seemed small before their visionary dividend'. Brunel was, he wrote, 'a plausible as well as a practical man'.

Work on construction of the line began in earnest in 1836, with successive reports to the shareholders noting progress at either end of the new railway. Brunel once more threw himself into the task ahead — the construction of the 116-mile railway he called 'The finest work in England'. Writing in his journal on Boxing Day 1835, he summed up the tumultuous events of the year in which he had emerged 'from obscurity'. He also noted the other 'irons in the fire' he had, namely the Clifton Bridge, the Bristol and Sunderland dock schemes, the Cheltenham & Great Western Railway, the Bristol & Exeter Railway, the Bristol & Gloucester Railway and the Newbury branch, which was 'a little go beneath my notice now'. The projects had, he calculated, a combined capital of over £5,000,000, not a bad total for a relatively inexperienced engineer of just 29. With these responsibilities also came a better salary; the £2,000 a year he received from the Great Western allowed him to move from cramped offices at 53 Parliament Street to a far grander property at 18 Duke Street, Westminster.

With better prospects, Brunel's thoughts also turned to marriage; there are numerous references throughout his diaries of encounters with the opposite sex, and there is little doubt that Isambard enjoyed the company of women, when he could find time in his hectic life. Writing in 1827 at the age of 21, he noted that he had, like most young men, 'had numerous attachments if they deserve that name'. After the ending in 1829 of a long relationship with Ellen Hulme, the daughter of a Manchester family whom he had known for some time, it was not until 1832 that (Brunel records) he visited the Horsley family in Kensington. It is likely that Isambard had been introduced to William and Elizabeth Horsley by his friend Benjamin Hawes, and he soon became a regular visitor to their house.

Of the Horsley children, John was to become an accomplished painter and member of the Royal Academy and, as a close friend of Brunel, was to paint his portrait on a number of occasions. As well as another son, Charles, the Horsleys had three daughters, Fanny, Sophy and Mary, and it was the latter who caught Isambard's eye from the first. Although she lacked the artistic gifts shared by her brothers and sisters, Mary was certainly the prettiest of the three sisters, although her cool beauty and seemingly disdainful manner led to the less charitable arguing that 'she had nothing to be proud of except her face'. It is likely that Brunel had determined to marry Mary early on in their relationship but, at the time he first met her, none of the projects he had worked on had yet given him enough confidence to propose to her. By the late spring of 1836, with construction of the Great Western underway and work progressing on the Clifton Bridge, Isambard felt ready to ask Mary to marry him. This he did, and the couple married on 5 July 1836 in Kensington Church. A two-week honeymoon followed, with Isambard and Mary travelling first to North Wales and then returning south along the Welsh Borders to the West Country. Brunel's greatest love, his work, was never very far away, and so it is not surprising that, during the trip, Charles Saunders met Isambard at Cheltenham to bring him news of progress on the new railway, and correspondence to look at.

When the couple returned to London, Mary settled down to life at Brunel's new home in Duke Street, giving birth to their first son, Isambard, on 18 May 1837. The younger Isambard was born with a deformed leg, which could have been cured through a simple operation, but his mother refused to allow surgery. Something of a disappointment to his father, Isambard Junior was sent away to boarding school at eight years old, which seems insensitive and harsh considering his disability. Although not sharing his father's interest or abilities in engineering, after Harrow and Oxford, the younger Isambard was eventually to become successful as a lawyer and as Chancellor of the Diocese of Ely. Mary was to bear Brunel two more children, Henry Marc, born in 1842, who went on to follow in his father's footsteps as a civil engineer, and a daughter, Florence, who married an Eton Master, Arthur James.

By August 1836 the GWR's directors were able to report that the first section of line, between London and Maidenhead, should be opened by the following autumn. By September 1837 most of the work had indeed been

completed; however, as well as the much-debated matter of the broad-gauge, Brunel had yet another innovation to introduce on the Great Western Railway. With gangs of navvies working hard to create the trackbed for the new line, Brunel had given great thought to the nature of the track itself. Once again, he was not content to copy the techniques used on other railways, but looked to an entirely new method of constructing the permanent way.

Other railways, such as the London & Birmingham, had used wrought-iron rails supported by cast-iron track-chairs fixed to stone blocks set into the ballast; Brunel preferred instead to introduce what later became known as the 'baulk road', namely rails supported by longitudinal timbers. Usually 30ft in length, these timbers were braced with cross-sleepers known as 'transoms', fixed at approximately 15ft intervals. This created a robust framework on to which the rail could be fixed. The whole structure was held down by 10in-diameter beech piles, which were driven into the ground between the rails, and bolted to each transom. Once the piles were driven in, ballast was tightly packed under the timbers to create a firm foundation for the track. By using this arrangement Brunel hoped to use a lighter rail than would be necessary when track was only supported at intervals by track-chairs. The rail itself was also of a new design, the inverted 'U' shape of the track section leading to its being called 'bridge rail'.

The adoption of this novel method caused the directors to abandon any hope of opening the railway in 1837; apart from the extra work required in actually laying the track, there were also delays in the delivery of both rail and timber to the railway. An additional problem was that the American pine used by Brunel for the baulk road needed to be treated to protect it. By 1840 creosote could be used, but before this became widely available, Brunel had to rely on a method called 'Kyanizing', where timber was dipped in chemical tanks. Defective and leaking tanks had to be replaced before adequate supplies of timber could become available. In its half-yearly report of February 1838 the company was forced to 'lament the unavoidable postponement of the opening', and also printed an extract of a report by Brunel on the track, in which he noted the method of construction in some detail.

Further difficulties were being experienced in the rather unorthodox locomotives specified by Brunel for his new line. Instead of using accepted standard designs of locomotive such as the Stephenson 'Patentee' type which was being manufactured for railways elsewhere, he wrote a general letter of specification to almost all the reputable locomotive builders in the country. No detailed plans or drawings were supplied, Brunel arguing that the

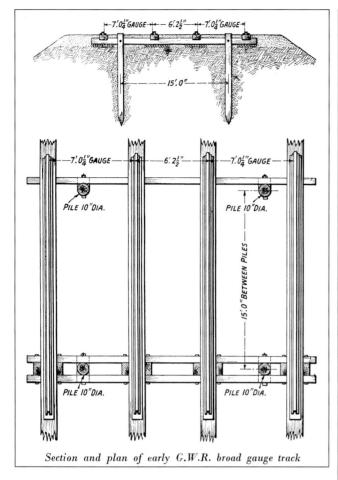

Section and plan of early G.W.R. broad gauge track

Above: Brunel's original broad-gauge 'Baulk Road' track.
Railway Gazette International

manufacturers should have the opportunity to introduce new improvements without the hindrance of detailed specifications. Prospective builders were told: 'The particular form and construction of the engines will be left to your own judgment.' Some parameters, stating the company's needs in terms of fuel consumption and speed of construction, were included. The few definite guidelines set by the engineer caused the companies the most problems. A minimum speed of 30mph was to be achieved without the piston-speed exceeding 280ft/min, and this should deliver a tractive effort of at least 800lb. Locomotives should, the letter insisted, weigh no more than 10½ tons.

Manufacturers duly produced locomotives to these guidelines, which, by their nature, led to engines with small boilers and cylinders, and large driving wheels. Writing in 1942, half a century after the end of the broad-gauge, one commentator argued that Brunel's locomotives were 'a collection of monstrosities that could barely shift their own dead weight'. When the Great Western board took the decisive step of appointing the 21-year-old Daniel Gooch

as Locomotive Superintendent in August 1837, no engine built to Brunel's specifications had yet been delivered to the railway, but as Gooch travelled around the various manufacturers where locomotives were being built, he can hardly have been impressed. Although very young, he had already had a varied career, working in South Wales, Scotland and the North East. He had worked at the Vulcan Foundry in Lancashire, and had been a draughtsman at Robert Stephenson's Forth Street Works in Newcastle between January and October 1836. He brought a sound and practical knowledge of steam locomotive technology, something he was to rely on in the coming years, as he struggled to keep Brunel's locomotive fleet running. Matters were improved when Gooch was able to secure the purchase of two 5ft 6in gauge engines from Stephenson, which had originally been destined for export to the United States. *North Star* and *Morning Star*, as the Great Western named them, were for a time the most reliable engines the company had.

The embarrassment caused by the delay in opening the line was nothing compared to the debate which followed the commencement of services on the London-Maidenhead section at the end of May 1837. Gibbs' diary for 20 June noted that: 'There has been some little disappointment as to our speed and the smoothness of our line.' The line had opened for paying passengers on 4 June, although the first trains had run on 31 May, when a Directors' Special, hauled by the locomotive *North Star*, had run from Paddington to Maidenhead, where a celebratory banquet for 300 was held in a tent.

Gibbs recorded that £226 was taken by the company on the opening day of service, but he was less interested in income, and more interested in the depressing effect the poor performance of the railway was having on the shares of the company. Brunel's new track was not performing well, and neither were the locomotives, most of which were extremely unreliable. Matters were not helped by the activities of a number of shareholders from the North of England, many of whom were opponents of both Brunel and the broad-gauge, and who now saw an opportunity to cause trouble. On 21 June Gibbs reported that Brunel had admitted that the line was 'in a very bad state', a situation caused by problems with the ballast. Pressure from the Liverpool shareholders to hold a special meeting was deflected by a promise to postpone the August half-yearly meeting by two weeks, to allow more time for a report on the situation to be considered. Gibbs, a staunch supporter of Brunel, was alarmed by reports that his removal was to be proposed at the August meeting. Despite travelling on the London & Birmingham Railway and finding the ride no better than the Great Western, Gibbs found the struggle with the Liverpool party was sapping his enthusiasm for the project. He would, he wrote on 7 August 1837, be sorry not to be involved with the building of the new line, but he would 'like still less to be controlled and dictated to by these Liverpool people'.

Having defended his cause during the Committee stage of both Great Western Bills, Brunel once again found himself fighting for the survival of his career. Not all members of the Board of Directors were as forthright in their support as Gibbs, and some wished to concede ground to the Liverpool party by allowing one of their number to join the board, and by appointing another engineer to work with Brunel. The half-yearly meeting held in Bristol on 15 August 1838 lasted for over seven hours, and Brunel was once again forced to justify himself in an eight-page report. With regard to the track he wrote: 'It may appear strange that I should again in this case disclaim having attempted anything new, yet regard to truth compels me to do so.' This method and the question of the broad-gauge were matters of principle, he wrote, which he would 'stand or fall by'.

Although Brunel was not called upon to resign, the directors were forced to concede that an independent engineer should be asked to report on the railway. This concession was less of a surprise to the board than might have been expected, since Brunel had already agreed to such a move in July, when three railway engineers, George Stephenson, Nicholas Wood and James Walker, had all been approached to report on the ailing railway. Of the three, only Wood had accepted, and even he could not produce any report until September at the earliest. In the meantime a further appointment was made, that of John Hawkshaw, engineer of the Manchester & Leeds Railway. It was Hawkshaw's report that was first to appear, but although bitterly critical of the whole idea of the broad-gauge as a system, it said little about one of the key issues it had been intended to tackle — the permanent way. The only comment made by Hawkshaw on the track was Brunel's technique of 'attempting to do that in a more difficult manner, which may be done at least as well in a simple and more economical manner'. The 27-year-old engineer's report was described by Gibbs as 'ill-natured',

Above: Sir Daniel Gooch, Locomotive Superintendent of the Great Western Railway, portrayed next to a model of one of his 'Firefly' class locomotives. *National Railway Museum*

Below: One of Brunel's 'freaks', *The Hurricane*, built by Hawthorne in 1838. It was not a conspicuous success, and had a short working life. *Author's Collection*

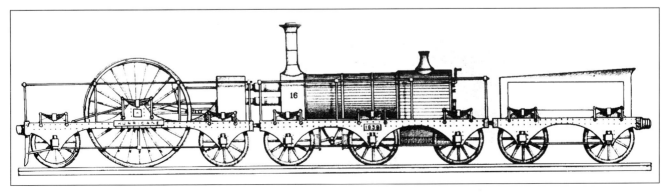

Above: The opening of the Great Western to Maidenhead, as recreated in 1935 for the GWR centenary film. The scene was re-enacted, using players from the GWR Operatic Society, on some waste ground at Swindon Works. *National Railway Museum*

Left: North Star, as rebuilt by the Great Western in 1854. The replica locomotive has been preserved at Swindon for many years. *Swindon Museum Service*

but to the Liverpool shareholders it only confirmed what they already believed. Some weeks later, Brunel replied to the criticisms levelled at him by Hawkshaw, particularly on the question of the broad-gauge. He could see little relevance in criticism that his railway, by its gauge, could not directly link with railways in other parts of the country, arguing that, since the Great Western and the companies associated with it were a complete system within the area they served, the whole question of interchange with other lines was an irrelevance. As McDermot notes, for someone as gifted as Brunel, this aberration with regard to the broad-gauge was short-sighted, and difficult to explain.

Meanwhile, work was continuing on the other report being compiled by Nicholas Wood, who had chosen as his assistant the redoubtable Dionysius Lardner. It was he who carried out a series of elaborate locomotive tests on the Great Western and a number of other railways. The results were not encouraging, Lardner finding that the Great Western locomotive *North Star* could haul 82 tons at 33mph, 33 tons at 37mph, and only 16 tons at 41mph. Moreover, to obtain the last result, coke consumption had shot up from 1.25lb/ton to 2.76lb/ton per mile. All this could be attributed to the larger profile of broad-gauge locomotives, Lardner argued, which increased atmospheric resistance. From this, Wood concluded, it would not be 'advisable to attempt an extreme rate of speed, and that 35 miles an hour...may be considered as the limit of practical speed for passenger trains'.

Brunel and Gooch soon realised that poor driving and firing had affected the results of the test, and further work on the blastpipe of the engine improved matters, allowing 50 tons to be hauled easily at 45mph. Wood's report, published on 12 December 1838, had also criticised the track construction, particularly the piles, which, he felt, did not assist in creating a firm base for the track. Increasing the size of the timber baulks would improve matters, he argued, and would still be better than stone block or timber cross-sleepers. On the question of the broad-gauge itself, Wood was somewhat ambivalent and, despite disagreeing with Hawkshaw's conclusions, seemed unable to come down firmly on the side of Brunel. However, the report was not well received by members of the Great Western board, some of whom suggested that Joseph Locke, Engineer of the Grand Junction Railway, should be brought in to work with Brunel. Offering to resign rather than accept this situation, Brunel was convinced that many of the conclusions in Wood's report could be confounded, given time.

Through the determination of Gibbs and some other members of the London committee, the whole question was postponed until after a special shareholders' meeting which was to be held on 9 January 1839. It was intended that this meeting should decide the question of the broad-gauge once and for all. The directors' report concluded that the shareholders should be 'deeply sensible of the disastrous consequences inevitably arising from the continual discussion of the principles acted upon in carrying of the works'. Brunel was able to show the much-improved performance of *North Star*, and cast doubt on much of Lardner's evidence in the process. With regard to the track, the directors' report accepted Wood's contention that the piles fixing the track were largely to blame for the rough ride experienced by passengers, and agreed to 'the fastening of the rails to the timber by screws'. A later writer argued that, as built by Brunel, the original track formation was 'a stiff and rigid road', and that 'in no appreciable sense elastic, it conformed to Brunel's view that elasticity in track was an evil'. The piles he used to fix the track down failed to do so, causing the rough riding he had been so careful to avoid in the first place.

'The experience of some months has now enabled the Directors to witness the progressive improvement of the working of the Railway,' the report optimistically noted. However not all the shareholders agreed, and a vote was held on the amendment: 'That the reports of Messrs Wood and Hawkshaw contain sufficient evidence that the plans and construction pursued by Mr Brunel are injudicious, expensive, and ineffectual for their professed objects, and therefore ought not to be proceeded with.' After much discussion, a vote was taken, and the next day it was revealed that Brunel's supporters had won the day, but only by a majority of 1,647. The engineer had one further indignity to suffer, that of further criticism over his choice of locomotives for the Great Western. The failure of many of the engines built to his specification prompted the directors to ask Daniel Gooch to prepare a report on the fleet as it then stood. Placed in such an invidious position, Gooch found his loyalty to Brunel compromised. Nevertheless, his report pulled few punches, and he was less than complimentary about the design and construction of many of the engines ordered for the railway. Not unsurprisingly, Brunel sent Gooch a sharp letter, but as Gooch was to note in his diary, although the tone of the letter was angry, 'he only shewed it in his letter, and was personally most kind and considerate to me, leaving me to deal with the stock as I thought best'. Having cleared the air, Brunel then allowed Gooch himself to produce designs for locomotives for the new line. The 'Firefly' class went on to become the mainstay of the fleet for some years, and most of Brunel's 'freaks' were quietly withdrawn and quickly forgotten. The whole question of the broad gauge, however, whilst settled for the present, had not been totally forgotten, and would return to trouble Brunel almost a decade later.

It was hoped that, with the Liverpool faction finally silenced, the completion of the rest of the Great Western Railway could be achieved with rather less trouble than encountered with the first, short section from London to Maidenhead. Over the next three years, the route would open section by section, with the line between Bath and Chippenham the last to be completed, in June 1841. Some mention was made, in the previous section of the book, of the long hours worked by Brunel; during the construction of the line, this situation continued, exacerbated no doubt by the way in which he worked. Although Brunel did employ a number of assistants, he was closely involved with almost every aspect of the huge project — a workload that is hard to imagine. From lamp-posts to large stations, the engineer made drawings, sketches and specifications, all of which he checked. He also dealt with contractors, landowners and all manner of other people, as well as producing the reports and information needed by the officers and directors of the company. Added to this was the responsibility for the *Great*

Western and, later, *Great Britain* steamships, which were built concurrently with the construction of the Great Western Railway. It is thus no surprise that Brunel's health would eventually suffer, although the strain of events during the construction and launch of the *Great Eastern* would take an especially heavy toll.

Rather than attempt a chronological survey of the construction phase, this chapter will instead take the form of a journey down the line from London to Bristol, describing some of the principal features on the route, as originally built. Although some attention was paid in Chapter 2 to the first section of the railway from London to Maidenhead in terms of the difficulties faced by the company in the first year of its opening, little has been said about the London terminus of the line, apart from the difficulties encountered in settling on a site for it, in the earliest years of the company's existence. With the abandonment of the London & Birmingham arrangement, the company chose another site for its station at Paddington. At the time Brunel was planning the Great Western, Paddington was a small village, some considerable distance from the centre of London, which then did not extend much further than Marble Arch. Within a few years, however, the place had changed dramatically. Writing in 1853, John Fisher Murray remarked that Paddington, 'where the traveller resigns

Left: The interior of Box Tunnel, from a lithograph by J. C. Bourne.
Swindon Museum Service

Below: The original Paddington station in around 1845.
Illustrated London News

Above: A rather odd bridge over a crossroads at Uxbridge only a few miles from the Paddington terminus of the Great Western Railway, from a lithograph by J. C. Bourne. *Swindon Museum Service*

Below: Ealing station c1845. *National Railway Museum*

himself to the locomotive agency of steam, like Islington, and many other suburban villages, has now ceased to form part and parcel of our environs, being absorbed in the far-extending town'. Since it was not Brunel's first choice of location for a station, it was built in a hurry, and was opened a few months after the opening of the first section of line, in 1838. Running across the east end of the station, beyond the buffer-stops, was Bishop's Road, supported on a series of arches which formed the entrance to the station itself. West of Bishop's Bridge, the facilities offered were limited to two arrival and two departure platforms, served by two traversers which allowed locomotives and carriages to be moved without extensive shunting. At the west end of the site there was a locomotive shed and workshops, where Daniel Gooch spent much of his time in the early years, attempting to keep the ailing locomotive fleet working.

The accommodation for passengers was far from adequate. In 1839 there were only 14 daily departures from the station, but as subsequent sections of the railway opened, and traffic increased, the facilities became less and less suitable. The opening of other railways such as the South Wales Railway and the Bristol & Exeter Railway, and arrangements with other lines, particularly in the Midlands, meant that there was considerable pressure to build a larger station. In the directors' report of the company's half-yearly meeting on 13 February 1851, it was stated that the board had considered that 'the time has arrived when a commencement must be made in the construction of more permanent buildings, fitted for the great increase in business'. It had been 'maturely considered', the report continued, 'whether any temporary additions to the existing buildings at Paddington could have been usefully made, so as to postpone the construction of a permanent station'. This, the directors concluded, would not have been economic, so the decision was made to start building a new station, with the sum of £50,000 allocated for work to start.

Brunel had always intended that the area to the east of Bishop's Bridge Road, bounded by Eastbourne Street and Praed Street, should be the site of a grander station, but lack of time and, more significantly, lack of capital prevented the construction of the kind of London terminus the Great Western deserved. It was more than 10 years before Brunel was able to write that: 'I am going to design, in a great hurry... a station after my own fancy; that is with engineering roofs etc. It is at Paddington.'

This new station would be no temporary timber affair; this one would be 'interior and roofed in', and in metal. The terminus was to be built on the site of an existing goods shed, and thus a good deal of inconvenience and disruption ensued from 1851 until the opening of the new station in 1854. In 1852 Brunel had reported to the shareholders that it had proved impractical to complete the departure platforms of the new station until the goods depot had been demolished and relocated. 'Extraordinary rain and floods' noted elsewhere in the half-yearly report had also 'retarded work very much for several weeks'. Some further idea of the difficulties encountered by Brunel may be deduced from his comments in early 1854, when he told shareholders: 'The difficulties of proceeding successively with different portions of the New Work on the site of the old buildings, without interfering too much with the carrying on of traffic in a station already too small for the wants, has been very great.' In the end, the departure platforms, on Eastbourne Terrace, opened for use in January 1854, with the arrival platforms opening five months later, in June.

Whatever the disruption, the end result was surely worth it. Brunel's new station consisted of three great roof spans: two smaller, of 69ft 6in and 68ft, and a larger central span of 102ft 6in. Each span was 700ft long, with wrought-iron ribs situated at 10ft intervals, which were supported by cast-iron columns. These columns, capped with elaborate capitals, were supplied by Fox, Henderson & Co, and were bolted into concrete bases, but were not reinforced with tie-rods across the spans. This arrangement was not a success, however, for when engineers checked the columns in 1915, prior to the erection of a new roof span, they found that the untied arches had pushed the whole structure out of line - so much so, that columns in the north span were out of line by 5½in. Eventually all the cast-iron columns were replaced by the GWR, a process not completed until 1924.

The design of the roof owed much to the work done by Joseph Paxton, inventor of the 'patent glazing system' used on the recently-opened Crystal Palace, and Rolt wrote that Brunel was 'a fervent admirer of Paxton's building'. Since the new station would essentially be set within a cutting, the roof was to be its highlight, and Brunel turned for assistance to Matthew Digby Wyatt, Secretary of the Executive Committee of the Great Exhibition, with whom he had worked during the preparations for that event. Wyatt had designed a 'Moorish Pavilion', which Brunel had much admired and so invited him to decorate Paddington. In typical fashion, Brunel noted that, as far as the decoration was concerned, it was 'a branch of architecture of which I am fond and of course believe myself fully competent for'; however, for the detail, he continued, 'I have neither the time or knowledge'. Pressing Digby Wyatt to accept, he asked him not to let the Great Exhibition work prevent him assisting, concluding that: 'You are an industrious man and night work would suit me best'! As Rolt notes, Digby Wyatt was probably just as busy as Brunel at this time, but all the same he accepted the offer, and his work could be clearly seen in the wrought-iron screens at the end of the train shed, and in cast-iron decoration used to relieve the rather plain roof-ribs. The completion of the new station marked a high point in the railway work done by Brunel, and finally gave him a fitting terminus to the 'Finest Work in England'.

When the Great Western opened fully for traffic in 1841, the passenger travelling west out of London on Brunel's main line would have seen little but open countryside. Some early travellers did not find the speed of railway travel conducive to viewing the landscape, and Murray noted that some of the views west of London would be interesting, 'did the rapidity with which we are impelled permit us to contemplate it'. The stations at Ealing and Hanwell, both opened in December 1838, served small settlements which were eventually swallowed up as London spread outwards. Seven miles out of Paddington, and just east of Hanwell station, was the Wharncliffe Viaduct, the first major engineering work on the line. Rather than build a massive embankment, Brunel designed the largest brick structure on the London-Bristol line — a 900ft bridge which crosses the Brent Valley on eight massive arches, each having a 70ft span. Work, under the supervision of contractors Grissell & Peto, began in February 1836 and was completed 14 months later, in April 1837. Foundations were driven down into the London clay, supporting twin tapering, stone-capped brick piers which have been described as 'Nubian' Egyptian in style. The rise of neo-classical architecture in the Victorian era also generated an interest in Egyptian designs, which, although not as popular as Greek and Roman styles,

influenced architects and designers such as Brunel, who had also used an Egyptian motif in designs for the Clifton Suspension Bridge. The dramatic and striking design was added to in the 1870s, when the viaduct was enlarged to accommodate two standard-gauge tracks in addition to the two broad-gauge ones originally provided. The viaduct was named after Lord Wharncliffe, Chairman of the House of Lords Committee dealing with the incorporation of the Great Western Bill, to mark his assistance in steering the Bill through Parliament. His noble Lordship's coat of arms can still be seen high on the bridge, recognising his contribution to the stormy passage of the Great Western Railway Act at Westminster.

When the first timetable issued by the Great Western was published in *The Times* newspaper on 2 June 1838, it was noticeable that it recorded Slough as a stopping place for GWR trains. This may have seemed rather surprising, bearing in mind the level of opposition the company had encountered from the nearby Eton College, and the fact that the act had not permitted the erection of a station at this place. In their report to the shareholders' meeting on 31 August 1837, the directors stated that the attitude of the college authorities would 'deprive the public of that accommodation to which they would otherwise be justly entitled'. The company would, the report continued, 'convey any passengers desirous of travelling to or from the neighbourhood of Slough', regretting that it could not provide 'the ordinary conveniences of a station' for the travelling public in the area.

The company directors were not the only people aware of the problems caused by the lack of a station at Slough. Early in 1838, memorials were presented to the college by the Mayor of Windsor and numerous local citizens. Their requests were rejected, the Provost of the college arguing that it was the duty of the college authorities 'to withhold their consent to such an application and claim the protection granted to them by the clauses contained in the

Below: The striking lines of the Wharncliffe Viaduct, near Hanwell. The bridge was extended when the line from London to Reading was quadrupled. *National Railway Museum*

Act of Parliament'. Approaches were made to the Great Western by local people requesting some kind of service to Slough, and in April 1838 Charles Saunders wrote to William Bonsey, a Slough solicitor, that 'the Directors will stop certain trains as they pass through Slough, for the purpose of allowing passengers to alight from their carriages or to join them'.

This defiance could not be tolerated by the college, which applied to the Court of Chancery for an injunction to stop what it considered to be a blatant snub to the spirit, if not the letter, of the Great Western Railway Act. Two days before the railway was to open, the case was dismissed with costs, and even the appeal was rejected, the Lord Chancellor ruling that the company was not actually doing anything expressly forbidden by the Act. Amazingly, while the appeal was still pending, the college had the cheek to request a special train to take Eton pupils to London, for the Coronation of Queen Victoria on 28 June!

Passengers were forced to put up with the inconvenient arrangement for almost two years, with the Great Western unable to influence the college. In July 1838 the company wrote to the college, offering to pay for two staff to be employed at Slough station to prevent Eton boys travelling without permission. Meetings were held

GREAT WESTERN RAILWAY.—The public are informed that this RAILWAY will be OPENED for the CONVEYANCE of PASSENGERS only between London, West Drayton, Slough, and Maidenhead station, on Monday, the 4th June. The following will be the times for the departure of trains each way, from London and from Maidenhead, (excepting on Sundays,) until further notice :—

Trains each way.

8 o'clock morning ; 4 o'clock afternoon.
9 o'clock ditto 5 o'clock ditto
10 o'clock ditto 6 o'clock ditto
12 o'clock noon 7 o'clock ditto

Trains on Sundays each way.

7 o'clock morning ; 5 o'clock afternoon.
8 o'clock ditto 6 o'clock ditto
9 o'clock ditto 7 o'clock ditto

Each train will take up or set down passengers at West Drayton and Slough.

Fares of Passengers.

	First Class.		Second Class.	
	Posting Carriage.	Passenger Coach.	Coach.	Open Carriage.
	s. d.	s. d.	s. d.	s. d.
Paddington Station to West Drayton	4 0	3 6	2 0	1 6
to Slough........	5 6	4 6	3 0	2 6
to Maidenhead ..	6 6	5 6	4 0	3 6

Notice is also given that on and after Monday, the 11th June, carriages and horses will be conveyed on the railway, and passengers and parcels booked for conveyance by coaches in connexion with the Railway Company to the west of England, including Stroud, Cheltenham, and Glocester, as well as to Oxford, Newbury, Reading, Henley, Marlow, Windsor, Uxbridge, and other contiguous places. By order of the Directors,

CHARLES A. SAUNDERS, } Secs.
THOMAS OSLER,

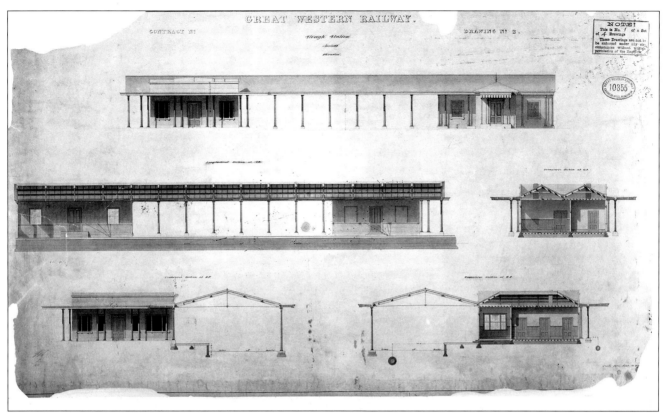

Above: Despite the protests of the Eton School authorities, a station was eventually erected at Slough, albeit later than anticipated. This Brunel drawing shows the station as originally planned. *Railtrack Great Western*

Above: The truly graceful lines of the Maidenhead Bridge are revealed in this Victorian lithograph. *Swindon Museum Service*

Above: Maidenhead Bridge in 1895, after its enlargement. The designs of the new extension to the bridge closely followed Brunel's own plans. *National Railway Museum*

between the two sides in 1838, although there appears to have been little common ground until 1839, when the college (its records reveal) was prepared to allow a station at Slough and also a branch line from Slough to Windsor. There appears to be little reason for the shift in position, although some pressure may have been forthcoming from the Royal Household at Windsor. In November 1839 Prince Albert made his first rail journey from Slough to London, and it seems that he and any other notable visitors to the Castle arriving by rail would have had to use the same temporary facilities as other travellers.

When the Great Western gave college authorities notice that it intended to repeal the parts of the Great Western Railway Act applying to Slough station, the college reply noted that although it had 'an insuperable objection' to the consent, it would 'take no step in opposition'. The station at Slough was finally opened in June 1840, built by Brunel in what became known as a 'one-sided' design. This rather peculiar arrangement consisted of two stations, with separate facilities for Bristol- and London-bound passengers. The two stations, each with their own awnings, platforms and buildings, were 50yd apart, with a complex web of sidings and wagon turntables in between. The biggest disadvantage of this system, which was also used at Reading, was that trains stopping at the platforms had to cross the main running lines. It seems likely that this rather

curious Brunel innovation lasted for almost 50 years, for it was not until 1886 that the station was finally rebuilt.

Pushing westwards across the Thames Valley, the railway seldom strayed far from the river, and at Maidenhead Brunel was faced with a considerable challenge in crossing the Thames, which at this point was over 100yd wide, and still a navigable channel. The river commissioners ruled that neither the towpath nor the channel should be obstructed, allowing Brunel only one bridge pier in the middle of the river. As Adrian Vaughan notes in his biography of Brunel, this stipulation is all the more ironic, bearing in mind the fact that, once the railway was operational, barge traffic on the Thames would all but disappear. The bridge had to allow sufficient clearance for barges, but Brunel was reluctant to raise the level of the trackbed, to avoid increasing the 1 in 1,320 ruling gradient from Paddington to Didcot. This situation led to the famous design, which had reputedly the flattest brick arches in the world. Two graceful, semi-elliptical arches of 128ft span the river, with smaller, more conventional arches on each side.

Above: Reading station in 1842, showing the 'double-sided' design. In the middle is a large goods shed. *Swindon Museum Service*

The bridge was built almost entirely of brick; writing some years later, Brunel's son speculated that, had the bridge been designed some years later, it might well have been built using cast iron or timber. As it was, the hundreds of tons of brick needed to complete this beautiful structure were to cause Brunel some problems in its construction. When the contractor expressed doubts that the design could work, Brunel, with his experience of the Thames Tunnel, was able to discuss with him the actual construction of the bridge, convincing him with diagrams and geometry that the whole structure would stand! His practical experience of the various trades involved in construction and engineering stood him in good stead, although much of the complex mathematics needed to design the arches was done by Brunel on a piece of foolscap paper, around his sketches for the bridge itself. The boldness of the design, however, unleashed the kind of criticism the engineer would have to endure throughout his career. As usual, Brunel did not let such criticism bother him, and work began on the bridge and the embankments approaching it in 1836. Despite the bankruptcy of the principal contractor, Bedborough, the following year, construction continued, and in 1838 the bridge was sufficiently complete for the replacement contractor, Chadwick, to ease the wooden centring away from the eastern arch. Brunel was ill at the time, and had not given permission for this to happen. Chadwick had been rather hasty, since the cement on the arch had not yet set, and as

a result the bottom three courses of brick moved by around half an inch. Brunel was understandably furious, and his critics saw this as a further opportunity to oppose his ideas, which were already under attack, particularly the whole question of his track and the broad-gauge.

What the critics could not ignore was the fact that the western arch was undamaged, and also Chadwick's admission in July 1838 that he had removed the centring too early, and would effect any repairs himself. These were completed in October without further difficulty, but the supporting scaffolding was not removed; perhaps as a joke at the expense of his critics, Brunel ordered that it should remain throughout the winter months, prolonging the calls by his opponents that it was the centring which was keeping the bridge from collapse. In November 1839 a severe storm led a rumour to circulate that the bridge was close to collapse. Great Western legend has it that during the storm the centrings were washed away, leaving the bridge still standing. However, more recent research has shown that the scaffolding was probably not taken down until early 1840, and only then under Brunel's instructions. When the Great Western main line was quadrupled in 1890 the bridge was widened using Brunel's designs, the enlarged structure reopening on 1 February 1893. No more graceful tribute than this could have been paid to the great architect, claimed the *Great Western Railway Magazine* in 1907.

Just east of Reading a hazard of another kind faced the great engineer. It had been intended to drive a tunnel of almost a mile in length through the hill at Holme Park near Sonning. However, it was stated in the directors' report of the third half-yearly meeting in February 1837 that arrangements had been made with the landowners for the conversion of the tunnel into an 'open cutting'. Although this change would entail additional expense, the directors thought that 'it cannot be doubted that considerable public benefit results from the change'. The new cutting was to be a much bigger affair, almost two miles in length, at a depth varying from 20ft to 60ft. Parliamentary permission for this and a number of other deviations from the original Act was sought, and granted over a year after the original public announcement.

The company may have regretted this change, as the cutting appeared to be beyond the ability of not one but two contractors. The first, William Ranger, was dismissed by the directors in August 1838, unable to complete this contract along with other work he had begun at the Bristol end of the line. Ranger simply could not cope with the sheer scale of the operation, and ran out of money. Thus began a legal battle which wound its way around the courts in Dickensian fashion, only ending in 1855 with a judgement in the House of Lords. Gibbs' diary for 28 August 1838 records that, by sub-letting the contract for Sonning Cutting, the work could now be completed in seven months. This approach seemed to be working, but only two months later, Knowles, the contractor responsible for the west end of the workings, was also dismissed, the navvies having gone on strike as he lacked the funds to pay them. Brunel and the company thus took over supervision of the work themselves. The shareholders were told: 'The Directors, finding that the only security for getting the earthwork completed in the course of the summer would be by taking it into their own hands, have done so.' The directors' report also shows the sheer scale of work needed to complete the railway designed by Brunel; over 1,200 navvies and nearly 200 horses were employed at Sonning, moving over 24,500cu yd of spoil in a single week. Conditions must have been very tough, even for these hardened souls, since conditions on site were never ideal. The cutting ran through a combination of clay, sand and gravel, which made the workings unstable, with landslips a common occurrence. Accidents were frequent, so much so that the directors gave a donation to the Royal Berkshire Hospital of 100 guineas, with an additional 10 guineas annually.

Although by August only 145,000cu yd still needed to be removed, the onset of another winter brought further problems. The storms referred to earlier in this chapter brought the workings at Sonning to a standstill, as the

Below: Sonning Cutting from a lithograph by J. C. Bourne. The timber viaduct in the foreground was the prototype for later bridges to be designed by Brunel in the West Country. *Swindon Museum Service*

heavy rain turned the excavations into a sea of mud. Brunel pressed on, and by the end of 1839 the cutting was finally completed. The 60ft cutting was crossed by two bridges, the more impressive being a three-arch brick structure which carried the main London to Bath road, whilst the other carried a minor road, and was probably the prototype for the larger timber viaducts Brunel was to build in the West Country some years later. Little of the spoil excavated from Sonning was wasted, most being used to construct embankments to the east of Reading. On 14 March 1840 the line to the capital was sufficiently complete for a special directors' train to be run, and a few weeks later, on 30 March 1840, the line was opened for traffic. The first train, the 06.00 from Paddington, was hauled by *Firefly*, the first Gooch-designed locomotive to be used on the Great Western, which had been delivered some weeks before.

The station at Reading was, like that at Slough, one of Brunel's 'one-sided' designs. Since the town was to the south of the line, Brunel built his station on the same side, and once again constructed two separate platforms, each with its own booking office and other facilities. In comparison to some of the grander stations on the line, Reading was built almost entirely of wood, and its shabby appearance was made worse when other buildings, such as a goods shed, were added. As at Slough, the station must have been a nightmare to operate, with trains crossing and recrossing the main line to gain access to the platforms. Brunel was somewhat relaxed about the arrangement, telling a Parliamentary Committee in 1841 that 'nothing but experience could determine whether a one-sided station was more safe or more dangerous than an ordinary station'. He evidently felt confident enough of the design to continue its use, for within a few years similar stations were built at Taunton, Exeter and Gloucester.

From Didcot, the Great Western main line continued down the Thames Valley, with the section from Reading to Steventon opening for traffic on 1 June 1840. Steventon

Below: One of Brunel's original drawings for a station on the Oxford branch. Named Dorchester Road on the plan, the station later became Culham. *Railtrack Great Western*

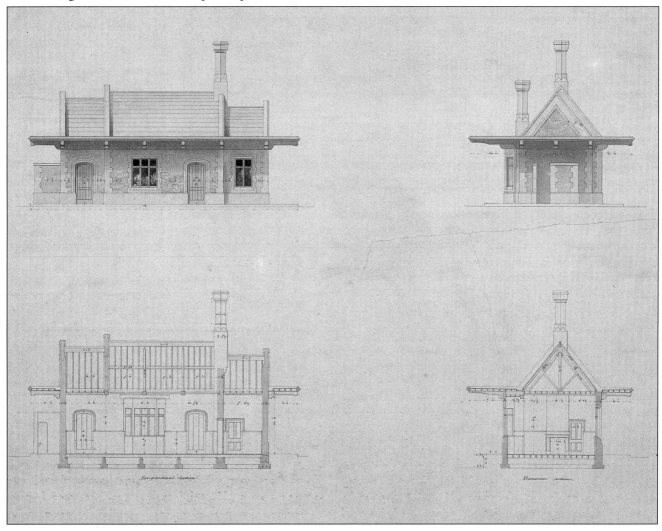

LONDON to BRISTOL.

On and after MONDAY, the 7th of December, 1840, the Line will be FURTHER EXTENDED to the **Wootton Bass[et]** **Road Station,** *(80 Miles from London,) for the Conveyance of Passengers, Carriages, Horses, Goods, and Parcels. T[he]* SHRIVENHAM STATION *will also be Opened.*

NOTICE.—London Time is kept at all the Stations, and will regulate the Arrivals and Departures throughout the Line.

LONDON TIME is   Minutes before READING Time.
Ditto   ditto   STEVENTON Time.
Ditto   ditto   SWINDON Time,
And the intermediate distances are in proportion thereto.

HOURS OF DEPARTURE AND TIME TABLE.

Down Trains, (Daily, excepting Sundays.)

DOWN / From Paddington	Departure from Paddington	Ealing	Hanwell	Southall	West Drayton	Slough	Maidenhead	Twyford	Reading	Pangbourne	Goring	Wallingford Road	Steventon	Faringdon Road	Shrivenham	Wootton Basset Road
To Wootton Basset Rd. A.M.	8. 0	..	..	..	..	8.35	..	..	9.13	9.26	9.34	..	9.59	10.17	10.34	10.50
— Maidenhead	8.30	8.41	8.45	8 50	8.59	9.10	9.20	..	..	..	..	..	..	..	..	..
— Wootton Basset Road	9. 0	..	..	..	..	9.35	..	10. 3	10.15	..	..	10 39	10.59	11.16	..	11.50
— Wootton Basset Road	10. 0	..	..	..	..	10.35	10.47	..	11.15	..	..	..	11.57	12.14	12.31	12.50
— Slough	10.30	10.41	10.45	10.50	10.59	11.10	..	..	..	..	..	..	..	..	..	..
— Wootton Basset Road	12. 0	..	..	..	12.26	12.38	12.50	1. 9	1.21	1.34	..	1.48	2. 8	2.26	..	3. 0
— Slough P.M.	1.30	1.41	1.45	1.50	1.59	2.10	..	..	..	..	..	..	..	..	..	..
— Wootton Basset Road	2. 0	..	..	..	..	2 35	2.47	3. 5	3.17	..	3.37	..	4. 1	4.19	4.36	4.53
— Wootton Basset Road	4. 0	..	..	..	..	4.35	4.47	..	5.15	5.28	..	5.42	6. 1	6.19	..	6.53
— Slough	4.30	4.41	4.45	4.50	4.59	5.10	..	..	..	..	..	..	..	..	..	..
— Reading	5. 0	..	5.13	..	5.26	5.28	5.49	6. 7	6.20	..	..	..	..	..	..	..
— Maidenhead	6. 0	6.11	6.15	6.20	6.29	6.40	6.50	..	..	..	..	..	..	..	..	..
— Wootton Basset Road	7. 0	..	..	..	..	7.35	..	..	8.13	8.25	8.35	8.42	9. 1	9.19	9.36	9.53
— Reading	8. 0	8.11	8.15	8.20	8.29	8.41	8.53	9.12	9.25	..	..	..	..	..	..	..
— Wootton Basset Road (Mail Train)	8.55	..	..	..	9.20	9.32	9.44	..	10.12	..	..	10.37	10.56	11.14	..	11.48

Sunday Down Trains.

From Paddington	Departure from Paddington	Ealing	Hanwell	Southall	West Drayton	Slough	Maidenhead	Twyford	Reading	Pangbourne	Goring	Wallingford Road	Steventon	Faringdon Road	Shrivenham	Wootton Basset Road
To Wootton Basset Rd. A.M.	8. 0	..	..	..	..	8.36	8.48	9. 7	9.19	9.32	9.39	9.46	10. 6	10.23	10.41	11.0
— Slough	8.30	8.41	8.46	8.52	9. 2	9.15	..	..	..	..	..	..	..	..	..	..
— Reading	9. 0	..	..	9.20	9.30	9.42	9.54	10.13	10.25	..	..	..	..	..	..	..
— Slough	9.30	9.42	9.48	..	10. 2	10.15	..	..	..	..	..	..	..	..	..	..
— Wootton Basset Rd. P.M.	2. 0	..	2.14	..	2.28	2.40	2.52	..	3.21	3.34	..	3.48	4. 8	4.25	..	5. 0
— Reading	5. 0	5.11	5.16	5.22	5.32	5.45	5.58	6.18	6.30	..	..	..	..	..	..	..
— Slough	7. 0	7.11	7.16	7.22	7.32	7.45	..	..	..	..	..	..	..	..	..	..
— Wootton Basset Road (Mail Train.)	8.55	..	..	..	9.20	9.32	9.44	..	10.12	..	..	10.37	10.56	11.13	..	11.48

Up Trains, (Daily, excepting Sundays.)

UP / To Paddington.	Wootton Basset Road	Shrivenham	Faringdon Road	Steventon	Wallingford Road	Goring	Pangbourne	Reading	Twyford	Maidenhead	Slough	West Drayton	Southall	Hanwell	Ealing	Paddington
from Wootton Basset Rd. A.M. (Mail Train.)	2.30	..	3. 1	3.18	3.36	..	..	4. 0	..	4.29	4.40	4.52	..	..	..	5.20
— Reading	..	..	..	..	..	..	..	7.30	7.39	7.56	8. 6	8.18	8.28	8.33	8.38	8.50
— Slough	..	..	..	..	..	..	..	..	..	..	9. 0	9.10	9.19	9.24	9.28	9.40
— Maidenhead	..	..	..	..	..	..	..	..	..	9.50	10. 0	10.10	10.19	10.24	10.28	11.26
— Wootton Basset Road	8.30	8.47	9. 3	9.19	9.37	9.46	9.54	10. 6	10.19	10.37	10.49	..	..	..	..	..
— Reading	..	..	..	..	..	..	..	11. 0	..	11.25	11.35	11.47	11.57	12. 3	12. 8	12.20
— Wootton Basset Road	10.15	10.32	10.48	11. 5	11.23	..	..	..	11.47	12. 0	12.27	..	..	..	..	1. 5
— Wootton Basset Road	11.30	..	12. 1	12.18	..	12.42	12.50	1. 2	12. 0	1.31	1.42	..	..	..	..	2.20
— Slough	..	..	..	..	..	..	..	..	..	..	3. 0	3.10	3.19	3.24	3.28	3.40
— Wootton Basset Rd. P.M.	1.15	1.32	1.49	2. 6	2 25	..	..	2.49	3. 3	..	3.31	..	..	..	..	4.10
— Wootton Basset Road	2.30	..	3. 2	3.19	..	..	3.49	4. 2	..	4.30	4.48	..	..	..	..	5.20
— Slough	..	..	..	..	..	..	..	..	..	..	6. 0	6.10	6.19	6.24	6.28	6.40
— Wootton Basset Road	4.30	4.47	5. 3	5.19	5.37	5.46	..	6. 4	6.17	6.35	..	..	..	..	..	7.20
— Maidenhead	..	..	..	..	..	..	..	..	..	7.45	7.54	8. 6	8.16	8.22	8.27	8.40
— Wootton Basset Road	6.30	..	7. 0	7.16	..	..	7.46	7.58	..	..	8.35	..	..	..	..	9.10

Sunday Up Trains.

To Paddington	Wootton Basset Road	Shrivenham	Faringdon Road	Steventon	Wallingford Road	Goring	Pangbourne	Reading	Twyford	Maidenhead	Slough	West Drayton	Southall	Hanwell	Ealing	Paddington
from Wootton Basset Rd. A.M. (Mail Train.)	2.30	..	3. 1	3.18	3.36	..	..	4. 0	..	4.29	4.40	4.52	..	..	..	5.20
— Reading	..	..	..	..	..	..	..	7.30	7.39	7.56	8. 6	8.18	8.28	8.33	8.38	8.50
— Slough	..	..	..	..	..	..	..	..	..	..	9. 0	9.10	9.19	9.24	9.28	9.40
— Wootton Basset Rd. P.M.	2. 0	2.17	2.34	2.49	3. 8	..	3.22	3.34	3.48	4. 7	4.18	..	..	..	..	5. 0
— Slough	..	..	..	..	..	..	..	..	..	..	5. 0	5.10	5.21	5.28	..	5.45
— Slough	..	..	5.32	5.47	6. 6	6.14	6.22	6.34	..	..	6.30	6.46	..	6.55	7. 1	7.15
— Wootton Basset Road	5. 0	..	5.34	5.49	6. 8	6.16	6.24	6.36	6.48	7. 8	7.28	7.34	..	..	..	8. 5

Above: A lithograph by one Mr Ann of Swindon of the scene looking to the west of the station in the 1840s. As well as the Works on the right of the picture, the Railway Village and St Mark's Church are visible.
Swindon Museum Service

Left: An early timetable for the Great Western Railway, showing the arrangements necessary whilst the line was not yet fully open.
Swindon Museum Service

was, until the opening of a station at Didcot in 1844, the principal station for Oxford, which could be reached easily from there by coach. The station at Steventon was of timber construction, but, close by, Brunel designed and built a fine house for the Superintendent of the Line. Subsequently this became the offices of the Great Western directors, being the site of board meetings until 1843. Built in Tudor style, much like Cirencester station, this fine building still stands today, although the station has long gone.

The line was extended to Faringdon Road on 20 July 1840, but it was not until the end of the year that the railway reached the small market town of Swindon. Even then, there were no facilities in the town itself, the line terminating three miles to the west at Hay Lane, not far from the village of Wootton Bassett. Today there is no trace of any station at the site, but for two years Hay Lane served as the principal station for Swindon, and although there are no known illustrations of it, a few sketches of the track layout survive in Brunel's notebooks.

Although a new station was to be built in Swindon, Brunel and Gooch had rather bigger plans for the town. For some time they had been discussing the construction of the company's principal engine shed and repair facility. Both Reading and Didcot appear to have been considered, as well as Hay Lane itself, but on 13 September 1840 Gooch wrote to Brunel, setting out his thoughts on the 'best site for our principal engine establishment'. This was to be at the junction between the Great Western Railway and the Cheltenham & Great Western Union Railway in Swindon, at this time an area of farmland, some distance away from the old market town itself. Swindon would be the place where locomotives could be changed, those with 7ft driving wheels used on the relatively flat Paddington-Swindon section being replaced by others with 6ft driving wheels for the more steeply-graded route from there to Bristol. The presence of the North Wiltshire Canal was also an advantage, Gooch wrote, 'communicating with the whole of England and by which we could get coal and coke'.

Only a lack of a suitable water supply was seen as a disadvantage by Gooch, and Brunel, agreeing with his Locomotive Superintendent, passed the report to the GWR board for approval. This was gained on 6 October 1840, and in the next half-yearly meeting of the shareholders, it was reported that the directors were to provide 'an Engine establishment at Swindon, commensurate with the wants of the company'. The establishment would also include 'large repairing shops for the Locomotive Department', the report continued, and 'this circumstance rendered it necessary to arrange for the building of cottages etc for the residence of many persons employed in the service of the Company'. The works was

Above: The First Class Refreshment Rooms at Swindon as reproduced in Measom's *Guide to the Great Western Railway* in 1852.
Swindon Museum Service

Left: The silver coffee pot from which the vile brew complained of by Brunel probably came. Constructed in the form of a 'Firefly' class locomotive, its appearance appears to have been more important than the taste of the coffee!
Swindon Museum Service

to be built on what would now be called a 'greenfield site', as already noted, some distance from the existing settlement of Swindon. Thus Brunel and Gooch created what became known as 'New Swindon' — a brand-new town consisting of railway works, station and houses. Gooch had noted in his letter to Brunel that the actual site at New Swindon should present few difficulties in the construction of the works; however, the greatest problem facing the company was how the complex would be paid for. There seems to have been no allowance for substantial repair facilities in the original estimates for the railway, but even worse was the fact that by 1840 the railway was well over its original budget of over £3 million. The result was that the Great Western was forced into entering an arrangement with the London contractors, J. & C. Rigby, who had also been engaged in the construction of Slough. In February 1841 the directors' report noted that the railway had entered into an agreement with 'responsible builders to erect refreshment rooms and cottages without the outlay of capital from the company'. The contractors would recoup this investment 'by passengers, consequent on the Trains stopping at that place'. It was also reported that Rigby's would receive rents from the cottages built for the company, although the Great Western would retain ownership of the land itself. The only capital needed, therefore, was to build the workshops at Swindon, which were, the directors emphasised, 'indispensably necessary'.

Before examining Brunel's role in the design of the works and Railway Village, it is worth starting our survey of New Swindon at the station, where the fullest ramifications of the agreement with J. & C. Rigby were apparent. This agreement contained a number of clauses which the Great Western would later regret. The first of these stipulated that all trains should stop at Swindon for ten minutes, to allow passengers to take refreshment. This may seem strange to today's Paddington-Bristol traveller, used to a journey time of around two hours, but in 1841 the journey could take up to four hours. As a further inducement to build the station, the company gave Rigby's a monopoly, allowing it to recoup its investment from the income from the Refreshment Rooms built there. It was also initially agreed that Swindon would be the only refreshment stop on the entire route. Brunel produced designs for the station buildings, which consisted of two 170ft blocks, linked by a bridge 'reminiscent', Rolt remarks, 'of the Bridge of Sighs', at first-floor level. Brunel also produced a specification for the station, on which the interior design was based. To be provided were 'Separate Refreshment Rooms, private rooms for ladies, water closets, urinals and lavatories for First and Second Class'.

Much attention was paid in the specification to the Refreshment Rooms themselves, such an important part

Below: Plan of Swindon Works in 1846. *Author's Collection*

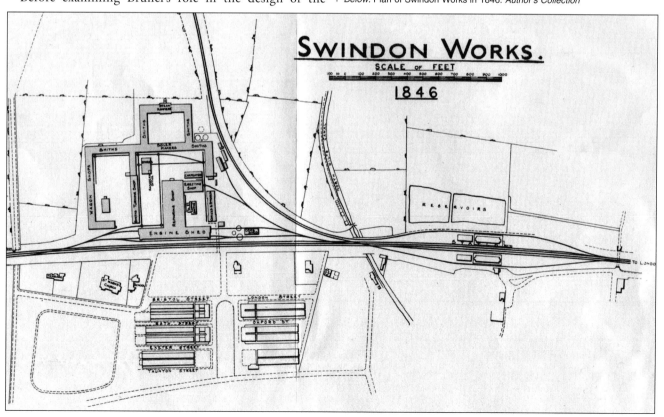

Above: The Engine House at Swindon in 1846. Thirty-six engines could be maintained at any one time within the building, and were moved around using the moving rail platform seen in the centre of the picture, known as a traverser. *Swindon Museum Service*

of the agreement with Rigby's. The interior of the Second Class rooms was 'to be distempered or papered above the dado, which, with other woodwork, shall be grained oak. As well as two large stoves, to keep travellers warm in winter, there was to be a ladies' waiting-room which was carpeted, unlike the main room, which was only covered in oil-cloth. Facilities for First Class were to be rather better; the rooms would 'be handsomely fitted up', Brunel wrote, and would have curtains and drapery to the windows, a carpet on the floor, and 'handsome oak chairs, tables and sideboards'.

On the first floor was the Queen's Royal Hotel, where passengers could break their journey and spend a night, albeit one punctuated by the sounds of passing trains! Quite what the facilities for guests were is difficult to determine, since there is little mention of what was provided. Downstairs, however, the reputation of the Refreshment Rooms spread far and wide very quickly.

Rigby's soon sub-let both the Refreshment Rooms and the Hotel to one Samuel Griffiths of the Queen's Hotel in Cheltenham, immediately recouping some of the outlay expended on construction of the station. Under new management, the quality of food and service in the Rooms quickly deteriorated. High prices and poor food gave the station the nickname of 'Swindleum', and Brunel himself was prompted to write, complaining about the coffee served and concluding: 'I seldom take anything at Swindon if I can help it.' There was little the company could do to extricate itself from the seemingly watertight agreement, and in 1848 the lease was sold by Rigby's to John Phillips, for £20,000. Despite various legal efforts to remove the 10 minute stop, which became unnecessary when locomotives were no longer changed at Swindon, the GWR was unable to break free from the lease, and only did so in 1895 by buying out the owners, for £100,000. By this time, having to stop all express trains on the main line was a huge handicap to the operation of the railway.

J. & C. Rigby were also the main contractors for the construction of the railway works, and work began in 1841. In their recent book on Swindon, J. Cattell and

K. Falconer noted that Brunel and Gooch probably designed the works together, with Brunel's architectural and design skills being balanced by Gooch's more practical background in heavy engineering. Two main tasks faced Brunel, in providing not only a large engineering and repair facility, but also a substantial engine shed, capable of stabling locomotives for the Cheltenham branch as well as for the London-Bristol main line. The engine shed was 490ft long, parallel with the main line, and at right angles to this was the engine house, used for stabling locomotives and carrying out light repairs. A traverser gave access to engine stalls on either side, while the most impressive structure was the 140ft-wide roof, constructed of wood tied with wrought-iron members.

The core of the works was concentrated in the area bounded by the Cheltenham branch line to the north, and

Above: A very early photograph of the Railway Works and Village, probably taken in the 1860s. The Wilts & Berks Canal can be seen in the foreground. *R. C. Nash*

Below: One of Brunel's original designs for the Railway Village at Swindon. *Railtrack Great Western*

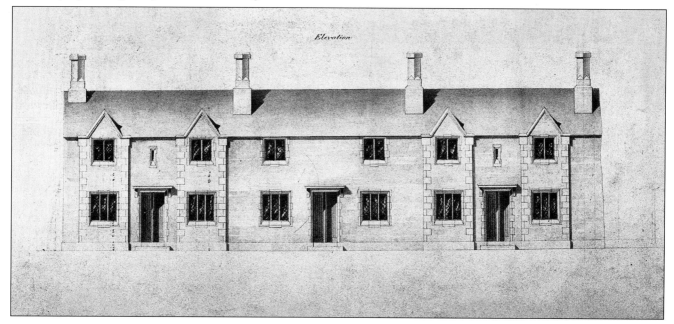

Above: Although heavily rebuilt over the years, the Bath stone used to construct part of Brunel's original Railway Works at Swindon in 1842-3 can be clearly seen in this modern photograph. Housing the office and general stores, this building formed part of the eastern boundary of the original complex. In the early years of this century another floor was added to the whole structure, housing a large drawing office. *Author's Collection*

Below: The Wootton Bassett Incline. This 1 in 100 gradient was the beginning of the steeply graded section between Swindon and Bath. On the left-hand side of the track, a Great Western signalman can be seen standing outside his box, awaiting the arrival of the next train in his section. *Swindon Museum Service*

the London-Bristol line to the south. The works began to grow in size almost before it was complete; recent research has shown that three distinct phases of construction actually took place in the first three years, so that by 1846, the year the first locomotive was actually constructed at Swindon, the works could boast a large engine house, boiler shop, carpenters' shop, machine shop, fitting shop and blacksmiths' shop. Brunel's designs for the works were rather plain, in comparison with some of the other buildings he built for the railway, but money was short, and perhaps Gooch's 'Quaker sensibilities', so disliked by Brunel, may have prevailed in this project, which was essentially a heavy engineering facility.

As the repair shops at Swindon were to be situated in a part of north Wiltshire with little tradition of heavy engineering, most of the labour required to build and run the new operation had to come from outside the local area. Examination of census data for the early years of New Swindon reveals that many of its residents came from areas already well-acquainted with railways: Scotland, the North East and the Manchester area. The company had already forecast that accommodation would have to be provided for members of the workforce and their families, and thus the agreement with J. & C. Rigby stated that the contractors 'shall at their own cost erect and build...many dwelling houses with or without shops or cottages not exceeding 300 in number, in such a manner as the said I. K. Brunel shall approve of.' The agreement also committed Rigby's to completing all 300 houses within a year, ready for use by the time the works came into operation in early 1843. This rather ambitious target proved impossible to achieve, and the contractors ran into financial difficulties, not least with the capital outlay necessary to build the station, works and cottages all at once! In the event, only 130 houses were completed by Rigby's, a year later than planned, at the end of 1843. The 300 cottages originally stipulated were not finally completed until 1855, four years before Brunel's death.

Contrary to much that has been written in the past, it seems likely that Brunel was the driving force behind the planning and design of the Railway Village. Writing to the architect Francis Thompson in January 1842, he noted that he was going to build cottages at Swindon of a 'totally unornamental character'. By March of the same year he had presented drawings of the first block of over 80 houses to the directors for approval, and work started on construction shortly after. Brunel appears to have had a hand in subsequent development in the Village, but it is probable that the drawings were produced by one of his assistants at his Duke Street office, with the engineer keeping an eye on progress. As noted earlier, progress was painfully slow, with the lack of accommodation for workers moving to Swindon leading to overcrowding and disease. The story of the Railway Village and its development has been well described elsewhere, so no further detail of subsequent events is necessary save the comment that, although Brunel may have considered his designs for cottages at Swindon 'unornamental' by the standards of industrial housing of the time, the houses provided for the workforce in the Village were of a very high standard, and the mixture of Jacobean, Gothic and Rustic designs have arguably left Swindon with an architectural gem.

The railway which ran west from Paddington along the Thames Valley to Swindon had been nicknamed 'Brunel's billiard table' because of its very shallow ruling gradient of 1 in 1320; although considerable work had been necessary to excavate the cutting at Sonning, the difficulties faced on this stretch of line were nothing compared to those experienced by Brunel west of Swindon. From Wootton Bassett to Bristol, extensive earthworks (in terms of cuttings and embankments) were required, as well as major tunnelling at Box, and between Bath and Bristol. The Hay Lane-Chippenham section contained various earthworks, all affected by the poor weather which continued to hamper the building of the line. In his report to the directors in February 1841, Brunel told shareholders that 'Exertions have been made in the whole of the last season to push forward the works...'; once again the project had been 'impeded by severe weather, severe and prolonged frosts, and subsequent floods'. One problem, exacerbated by the poor weather, arose when spoil, dug out by the navvies in the excavation of cuttings, was dumped wet to form embankments. This caused a number of landslips, and Brunel ordered that, wherever possible, 'the principle of avoiding the formation of earthwork in wet weather should be adhered to'. The problem was not entirely solved, and landslips continued to give trouble for some considerable time. In any event, problems on the Hay Lane-Chippenham section were rectified, and the line opened for passengers on 31 May 1841.

Beyond Chippenham, Brunel faced his greatest challenge. McDermot records that, between there and Bath: 'On scarcely one of the thirteen miles were the rails within ten feet of the natural ground.' Leaving the station at Chippenham, the line runs for almost two miles along an embankment, and then plunges into a deep cutting of almost three miles in length, before reaching perhaps the most impressive engineering achievement on the line, Box Tunnel. Rising 400ft above the level of the line, the hill consists mainly of oolitic limestone with layers of fuller's earth. Well before work started, there had been much debate about whether such a tunnel was feasible and, if it was built, how safe it would be. The questions first raised in the

Above: The approach to Chippenham station. The station itself can be seen in the middle distance. *Swindon Museum Service*

Committee stage of the Great Western Railway Bill in 1834 and 1835 continued unabated, forcing the directors to issue a statement in their half-yearly report to the shareholders in February 1837. Trial shafts had been sunk at Box, they reported, and the results obtained 'gave full assurance of the work being free from unexpected difficulties'.

In all six permanent and two temporary shafts were sunk vertically through the limestone strata, these being complete by the autumn of 1837. Contracts could be placed for the building of the tunnel itself, and in early 1838 George Burge, a contractor from Herne Bay in Kent, was appointed to excavate all but half a mile of the tunnel. The remaining section, at the west end, was to be excavated by two local contractors, Lewis and Brewer. Much of the tunnel was to be lined with brick, save at the west end where the bare rock would form the structure itself. The ambitious target of completion by August 1840 was set, a period of only 2½ years. The 3,212yd tunnel was for some years the longest in England, and the achievement is all the more impressive when it is realised that the tunnel was excavated largely by human labour. It has already been

noted that the navvies employed on the Great Western must have been a tough breed, but in the digging of the tunnel they showed remarkable resilience and tenacity. Gunpowder was used to blast the rock, but all other work was done by hand, with horses used both to haul spoil away, and to operate winches to hoist material up the shafts.

The presence of some 3,000 navvies in the neighbourhood caused much apprehension amongst local

Right: One of the slips given to navvies, allowing them to be issued with beer. Many unscrupulous contractors paid the navvies using the 'truck system', whereby they could redeem their pay only through the contractors' suppliers. *Author's Collection*

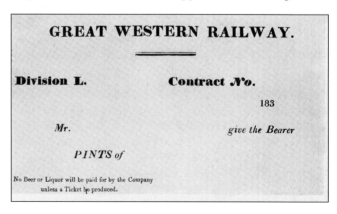

GREAT WESTERN RAILWAY.

Division L.   Contract №.

                                              183

Mr.   *give the Bearer*

PINTS of

No Beer or Liquor will be paid for by the Company
unless a Ticket be produced.

Above: Box Tunnel in later steam days. Smoke from previous trains still hangs in the entrance to the west portal. *National Railway Museum*

Above: Box Tunnel from one of Brunel's original engineering drawings, showing the west portal. These drawings are still in the care of Railtrack Great Western, which maintains Brunel's structure today. *Railtrack Great Western*

people. Most of the workers were housed in Box and Corsham, but, as contemporary sources note, the men worked continuous day and night shifts, so that 'as soon as one lot turned out, so another lot turned in, so that beds were never empty'. Drunkenness was rife, and was part of the navvy lifestyle. In 1850 it was calculated that the amount spent on alcohol per mile of railway in Great Britain was £100. There was also a good deal of fighting, although recent research on local court records has shown that, in many cases, it was the local population which had provoked fighting, not the navvies, most of whom could ill afford to lose their jobs. The local economy also boomed as a result of the railway works, many of the 300 horses used on the contract being supplied locally.

Work on the tunnel was hampered by the penetration of water into the excavations, and extra pumps were required. Adrian Vaughan records that Brunel was a particularly hard taskmaster in dealing with the contractor Burge who, struggling with water and beds of clay within the tunnel, did not make the sort of progress the engineer required. By August 1840 Brunel was able to report to the shareholders that 'five-sixths' of the tunnel had been completed, and that the tunnel might be finished by the next half-yearly meeting in February 1841. This prediction proved to be over-optimistic, and the works were not completed until 30 June 1841, after extra men and materials had been brought in to complete the job. Even then, only one line through the tunnel was opened for traffic. Gooch had to act as pilotman for every train passing through the tunnel, a task which he performed continuously for the first 48 hours. He recorded in his diaries that 'I was not sorry to get home and to bed at Paddington, after two days' and nights' pretty hard work'. This duty had not been without incident, as on the second night only Gooch's quick reactions had avoided a major crash, when two trains were inadvertently directed on to the same line by a sleepy signalman at the top end of the tunnel.

Gooch also noted in his diaries that, for the period when only one track was open, the candles of the men working to complete the line gave 'a perfect illumination, extended through the whole tunnel, nearly two miles long'. The whole question of lighting had already been an issue, and it had been planned to light the whole tunnel, in order to alleviate public anxiety about travelling through it in total darkness; some passengers of a nervous disposition would leave the train at one end of the tunnel and travel by coach over the hill to rejoin the Great Western at the other end. Brunel wrote to the directors, asking to abandon the idea, noting quite logically that the tunnel was no darker than the rest of the line was during the night, and that providing lighting might 'remedy the evil of darkness in tunnels', but would be 'very costly'.

There was some doubt as to whether excavations from either end of the tunnel would meet, but it was recorded that the tunnel roof was accurate 'to a hair', and that the side walls met within an inch and a half. All this was achieved without the satellite-guided machinery used by tunnellers today and, furthermore, the headings were driven to achieve a gradient of 1 in 100 through the hill. Since the opening of the tunnel, numerous stories have circulated regarding the penetration of the sun's rays through the tunnel each year on 9 April, Brunel's birthday. As early as 1850 the *Illustrated London News* recorded that this had occurred, but more recent research has shown that, although the sun has shone through the tunnel on a number of occasions, it was not on Brunel's birthday but on a number of other days in April, and once in September. The apocryphal nature of this Brunel legend can also be judged by the fact that few biographies of the engineer, including the one written by his son, even mention the story!

Although the completion of Box Tunnel had delayed the opening of the line in the spring of 1841, there were also problems nearer Bath. In February 1841 Brunel had reported that: 'The Diversion of the Kennet and Avon Canal in the progress of which very serious difficulties had occurred, principally by continued wet weather at a critical period of the works requires most attention.' Brunel was required to design a massive retaining wall to hold back the canal, which passed through Sydney Gardens, a Bath pleasure park. Over 27ft high, this wall was 5ft thick at its base, and formed a barrier with the canal, which had been relocated in land sold to the Kennet & Avon company in 1839. Brunel also took great pains with the design of the railway as it passed through Sydney Gardens, described by J. C. Bourne as the 'Vauxhall' of Bath. The architecture of the railway, Bourne continued, had been arranged 'to increase, rather than injure, the attractions of the gardens'. Two bridges linked either side of the park, and on the west side of the railway a low wall pierced by balustrades completed the effect.

Further engineering work was needed as the railway passed through Bath. The line runs through the city largely on embankments, crossing the River Avon twice. The original station, built in the style of 'James the First with debased Gothic windows and Romanesque ornaments', as Bourne described it, had an overall roof with a span of 60ft. Built on an embankment, the booking office and other facilities were in the basement, with the platforms at first-floor level, much as they are today, although the roof has long since gone.

The comparatively short section of line between Bath and Bristol contained further engineering difficulties. There were seven tunnels, numerous deep cuttings and a

Above: The railway at Sydney Gardens, Bath. *Swindon Museum Service*

Below: One of Brunel's original drawings for the station buildings at Bath. *Railtrack Great Western*

two-mile embankment near Keynsham. Other problems to be overcome included the diversion of a short section of the River Avon near Foxes Wood Tunnel, and the building of a 28-arch viaduct at Twerton, on the outskirts of Bath. The first contract for works in the Bristol area was placed with William Ranger in March 1836, but, despite good progress being made initially, difficulties with Ranger led to his removal from the contracts in the spring of 1838. The completion of works was, like progress elsewhere, restricted by the weather, and in February 1840 Brunel reported that flooding on the River Avon had caused much disruption, particularly in the construction of bridges, which had to be abandoned for some time, and also at the site of the new station in Bristol, which was itself flooded.

McDermot records that, in order to try and make up for lost time, the company was forced to work on the line night and day, and also on Sundays, much to the dismay of the citizens of Bath. These efforts were rewarded, however, when, after the completion of much of the work and the laying of the permanent way, the directors announced that the line from Bristol to Bath would open on 31 August 1840. Ten days before, some of the directors of the Bristol committee enjoyed their first taste of railway travel when they made the first journey between the two cities. Accompanied by Brunel, the five directors, Clarke, the Superintendent, and 'other officers' of the company all crammed on to the footplate of the 'Firefly' class locomotive *Arrow*. No carriages were provided, since the rails into the station were not yet complete, so the party

Great Western Railway tunnel near Bristol, 1840.

Above left: A timber skew bridge built to the west of Bath station. It was recorded that no contractor was interested in taking the contract and Brunel was forced to do much of the design work himself at great speed. Overall, the effect was very pleasing, and the bridge was not replaced by the Great Western until 1878. *Swindon Museum Service*

Left: One of the tunnel entrances on the Bristol-Bath stretch of line. This example is vaguely modelled on a Norman theme. Others had turrets and battlements like medieval castles. *National Railway Museum*

Above: Keynsham station, between Bristol and Bath, in standard-gauge days.

travelled on the locomotive itself, arriving in Bath 33 minutes later. As the locomotive 'flew onwards', a contemporary newspaper recorded, 'the party were [sic] greeted with hearty cheers from bands of workmen and spectators at various points'. When the railway opened for paying passengers there was little pomp and ceremony, but the running of the first train did cause much excitement in the area, with over 5,000 passengers carried on the first day.

Bourne, in his portrait of the Great Western Railway, argued that the new station at Bristol was a building of 'considerable pretension'. Situated at Temple Meads, some way out of the city centre, the station was built in the style of an Elizabethan manor house, with mullioned windows. There is some evidence that members of the Bristol committee felt that the architectural features of the station should be in harmony with others within the ancient city, whatever the cost — not something the seemingly more prudent London committee would appreciate. In July 1839 Thomas Osler, Secretary of the Bristol committee, wrote to Charles Saunders, Secretary of the London board, arguing that Brunel's 'Tudor' design was nevertheless only £90 more than a design he called 'as thoroughly naked an assemblage of walls and windows as might be permitted to enclose any Union Poor House in the Country'. Presumably the London committee agreed, as the station was more or less designed as shown in the original plans.

The train shed itself was situated at first-floor level, since the railway was actually 15ft above ground level at this point. The castellated frontage facing Temple Gate contained a boardroom and offices; passengers entered the station at ground level and, having bought a ticket, proceeded upstairs to the platform. The most impressive feature of the station was the mock-hammerbeam roof, built entirely of wood in the manner of Westminster Hall in London. Five broad-gauge tracks were accommodated without obstruction under the 72ft span, supported on either side by cast-iron colonnades. At the far end of the station, facilities were provided for the maintenance of locomotives, with chimneys to remove steam and smoke. A large water tank in the roof also served the station, and a goods depot was provided nearby. However, the station soon became too small for the increasing traffic on the Great Western, and it was extended in 1865, when a far larger station, providing facilities for the GWR, Bristol & Exeter and Midland, was built.

The opening of the complete Great Western main line on 30 June 1841 was marked with little ceremony; a Directors' Special left Paddington at 08.00, arriving in Bristol four hours later. Perhaps by now the novelty of railways in this part of the country had worn off; Brunel had cause to celebrate the opening of 'The Finest Work in England', but he had many other things to worry about — the extension of his railway far beyond the existing London-Bristol line, and farther still, across the oceans, using steamships to create a greater Great Western. There is little doubt, however, that the completion of this great work took its toll on the engineer; throughout the planning and construction of the railway, Brunel worked constantly, travelling up and down the route in his 'Flying Hearse' carriage, snatching what sleep he could, when he could. For much of the time, he worked for 20 hours at a stretch; his wife, Mary, saw very little of him, and could often only expect him home at weekends. Writing to her when working at the Bristol end of the line, Brunel reported that he was settled in a Wootton Bassett inn for the night, having walked 18 miles from Bathford during the day. After the completion of the Great Western, the first signs of ill health began to show, and in 1842 he was confined to bed for 10 days, and unable to attend a board meeting of the company. No doubt the effects on Brunel's health were cumulative, not helped by the filthy and insanitary conditions he had encountered during his work on the Thames Tunnel. The engineer was soon back to work at the old hectic pace, but there is evidence that, by the end of the decade, even he had begun to consider slowing down.

Below: The railway between Bath and Bristol does not stray far from the River Avon. This lithograph shows the line near Foxes Wood Tunnel. *Swindon Museum Service*

Above right: The imposing frontage of Brunel's Bristol terminus, from a drawing by S. G. Tovey. Passengers were taken by carriage through the arch on the left of the building, where they then proceeded up stairs to the platforms. *Swindon Museum Service*

Below right: The interior of the Bristol terminus. Locomotives were moved from one side of the station to the other by the use of a traversing table, which can be seen in the foreground. *Swindon Museum Service*

Chapter Four

The Bristol Steamships

Some idea of Brunel's enormous capacity for work may be gathered from the fact that, as well as acting as the architect, designer and engineer of the Great Western Railway from its inception to completion in 1841, he was at the same time engaged on another project — the construction of not one, but two transatlantic steamships. Although he was eventually to design a third and much larger vessel, the *Great Eastern*, he began more modestly with the construction of the *Great Western*, launched at Bristol in 1838. Brunel's son Isambard described the occasion on which his father first proposed this new venture. It was at an early meeting of the Great Western Railway directors, held at Radley's Hotel in Blackfriars. One of the directors commented on the enormous length of the railway as it was then planned, and Brunel is said to have retorted: 'Why not make it longer, and have a steamboat to go from Bristol to New York, and call it the *Great Western* ?'

Not surprisingly, most of the assembled directors did not take Brunel's comment very seriously; however, one, Thomas Guppy, did not treat the suggestion as a joke, and having discussed the matter further with him, determined to set up a committee to bring the project into being. Guppy, who was both an engineer and a businessman, soon persuaded three further GWR board members, Robert Bright, Thomas Pycroft and Robert Scott, to join this committee. Another recruit was Capt Christopher Claxton, RN, an old acquaintance of Brunel, whom he had first met in 1832, when working on proposed improvements to Bristol Docks. The next task was to raise enough money to build the ship. Such was the fear of the scheme being copied by others, or of speculators moving in, that only six copies of the first prospectus were issued, and these were handwritten, not printed. There was no initial advertising, save word-of-mouth canvassing by local businessmen. The prospectus proposed that the company should build two 1,200-ton steamships with 400hp engines, and also stated that no meeting of the shareholders could be held until at least half the shares had been allocated. Despite the somewhat clandestine way in which the prospectus was circulated, by March 1836, the date of the first public meeting of subscribers to the scheme, over 1,500 out of 2,500 shares had been allocated. Many shares were purchased by local people, although Brunel did invest in the project himself. In the hiatus before this public meeting, Claxton, Guppy and William Patterson, a Bristol shipbuilder of some reputation, toured a number of shipbuilding centres, including Glasgow and Liverpool. Using his contacts in the Royal Navy, Claxton

Opposite: The launch of the *Great Western* on 19 July 1837. As the picture shows, large crowds gathered for the event, many crowding on to ships in the harbour itself. The large church tower in the background is that of St Mary's Redcliffe. *City of Bristol Museum & Art Gallery*

was able to gain access to the Naval Dockyard at Woolwich, and to Admiralty drawings of ships currently under development.

It may be useful at this point to put Brunel's steamship scheme into a broader context. Although very primitive steamships were developed in the latter part of the 18th century, it was not until the second decade of the 19th century that steamships began to be used in any numbers. Many of the earliest were river craft, either tugs or pleasure steamers, only rarely venturing into salt water. In 1812, for example, the *Comet* was employed on the run between Glasgow and the Scottish Highlands, and two years later it was reported that steamers were regularly in use between London and Margate. The first cross-channel steamship was the 112-ton *Hibernia,* which tackled the Holyhead-Dublin crossing in seven hours in 1816. Development then quickened, and C. R. Gibbs notes that in 1825 there were 44 steamers under construction in the ports of London and Liverpool alone. Most of these ships were, however, primarily designed to steam in the shallower waters around the coast although, given favourable weather, they could venture further.

The early steamships also had a limited range, which was due in no small part to the lack of space available in their bunkers for coal. No better illustration of this can be given than that of the first transatlantic crossing by a steamship, which took place in May 1819. The *Savannah*, a 350-ton coaster which Denis Griffiths describes as 'little more than a square-rigged sailing ship with a small steam engine', crossed the Atlantic in search of a purchaser. The passage from Savannah, Georgia, to Liverpool took 29 days 11 hours. On only a dozen occasions during the trip were the engines used for more than 10 hours at a stretch; for most of the voyage, the ship relied on its sails, the paddle wheels being lifted out of the water. No passengers or cargo were carried, so the trip cannot really be seen as a serious milestone in the transatlantic steamship saga. More significant were the voyages of the *Curacao*, a Dutch-owned ship which made a number of crossings between Holland and West Africa in 1827 and 1829.

What the exploits of these small vessels did show was that, to be successful, transatlantic steamships would need substantially larger bunkers, since early vessels used huge quantities of coal. Stronger and substantially larger hulls were also needed, not only to withstand the battering that North Atlantic storms could unleash, but also to house the far bigger engines and boilers needed to complete a crossing. With these design considerations it was obvious that a ship far larger than anything built to date would be required. Commercial concerns also meant that only a ship of around 1,000 tons would be large enough to house the larger engines and coal bunkers already mentioned, and provide space for

Above: The facade of what was originally named the Royal Western Hotel, designed by R. S. Pope in collaboration with Brunel, and completed in 1839. The hotel was intended as a resting place for passengers travelling from London by the Great Western railway where they could pause before embarking for the United States on the *Great Western* steamship. The hotel closed in 1855, and for a period housed Turkish baths. Today only the facade remains, modern offices having been built behind. *Author's Collection*

cargo and passengers, sufficient to generate revenue to justify its existence.

To many, the prospect of such a ship ever being successful was non-existent. This belief was based on the premise that, to increase the size of a ship to accommodate more coal, its engine capacity would have to be increased by the same proportion. Thus, argued the sceptics, doubling the size of a ship would also mean doubling its engine size. Dionysius Lardner, Brunel's old sparring partner, went so far as to exclaim in 1835 that, in relation to transatlantic steamship travel, people 'might as well talk of making a voyage from Liverpool or New York to the Moon'. This comment was to attract much attention, and most observers were less than complimentary, although there is evidence that his claims may have discouraged some from investing in the Great Western Steamship Company. Within a year, Lardner had been forced to water down his rather rash comments, perhaps as a result of some of Brunel's own thoughts on the matter. In the report issued by the Great Western Steamship Committee in January 1836, Brunel demonstrated that the premise noted above was false. The construction of a larger vessel, he argued, would increase the carrying capacity of the hull in cubic yards, whereas the surface of the hull in contact with the water only increased by square yards. Vaughan notes

that even Brunel conceded that this idea was not his own, being already common knowledge amongst shipbuilders, but Brunel's achievement was in putting the theory into practice, to build some of the finest steamships ever constructed.

A new prospectus was issued just before the first meeting of the Great Western Steamship Company on 3 March 1836. This gave further details of the two ships to be built, their cost, and the likely return on investment. The cost was estimated at £35,000 each, and, if they completed the proposed 12 crossings of the Atlantic each year, it was hoped that a dividend of 15% could be paid to investors. At the meeting, a board of directors was appointed, which included such familiar names as Robert Bright, Christopher Claxton and Thomas Guppy. Not surprisingly, five members of the board were also board members of the Great Western Railway. The company itself, which did not actually come into being until June 1836, was what was known as a 'large partnership' rather than a public company, which obviated the legal requirement for public meetings. Brunel and the company were thus able to avoid a damaging public debate as the project progressed, a situation he unfortunately could not avoid with the Great Western Railway.

Although Brunel was largely responsible for the design of the ships, he requested that a 'Building Committee' be set up, which included Brunel, Guppy and Claxton. Also on the committee was William Patterson, who, with his partner, John Mercer, was a well-known Bristol shipbuilder. Patterson was known as 'a man open to conviction and not prejudiced in favour of either quaint or old-fashioned notions in shipbuilding', qualities which, no doubt, made him an ideal candidate in Brunel's eyes. It is recorded that work on the Great Western Railway brought Brunel to Bristol at least once a week at this time, and the committee thus met either at the shipyard office or at Claxton's or Guppy's house. Isambard's son wrote that 'they often sat far into the night discussing the details of the ship'. The keel of the new steamship was laid without ceremony in Patterson's yard at Wapping, in Bristol's Floating Harbour, in June 1836. Brunel had hoped to increase the size of the ship before construction began, but on the advice of Patterson did not do so, instead leaving changes to the second vessel. Even so, at 205ft, the keel was the longest that had been laid down to date. Much attention was paid to the strength of the hull, in order that it should withstand the stormy Atlantic seas, and Brunel was able to report that the ship was 'most firmly trussed with iron and wooden diagonals', and that the whole structure was held together with screws and nuts, 'to a much greater extent than has hitherto been put to practice'.

The building of the enormous ship attracted great attention in Bristol and, despite the quiet start to construction, thousands turned out to witness a short ceremony when the stern-post of the ship was raised, on 28 August 1836. Cannons were fired, and the stern-frame placed in position. The directors and invited guests then moved to the shipyard's mould loft, where a cold collation was taken, and numerous congratulatory speeches made. The sheer size of the ship had led many to question whether it could be successfully launched and navigated out of the dock complex along the very narrow channel of the River Avon to the Severn Estuary and the Atlantic. Claxton in his speech laughed off such worries, joking that 'the ship is to be made with a joint in the middle'. The size of the ship would, however, cause its builders some problems, which would be repeated with the launching of the larger *Great Britain* in 1843.

The irrepressible Dionysius Lardner paid another visit to Bristol less than a month later, to give a lecture to the British Association for the Advancement of Science. Although his pronouncements were rather less dramatic than previously, he nevertheless still felt that Atlantic steamship travel experiments should be carried out with 'great caution'. Using data he had collected over some years, he concluded that the maximum range of a steamer would be 2,080 miles. Although the exact details of the meeting have not survived, it was reported that Brunel, whilst pointing out errors in Lardner's calculations, failed to win over many of those present, who were impressed by the lecturer's 'dogmatic assertions'. The notion that steamships might have a limited range was eagerly taken up by many in Ireland, and one entrepreneur planning a railway from Limerick to Waterford was moved to write that the promoters of Atlantic travel between England and New York 'stand forewarned of defeat'. The public should be comforted with the thought, he continued, that 'as the exhausted mariner returns, he will fall in with the western shores of Ireland' and, reaching the coast, exclaim: "'This is the place from which I ought to have set out, for here I have returned with ease and safety!"'.'

Brunel and the Great Western Steamship Co were evidently unimpressed by such hyperbole, and construction continued apace. Tenders were invited for the supply of engines. Three were considered, and after consideration, the London firm of Maudslay, Son & Field was chosen. A local company, Winwood's, was one of the two unsuccessful bids, and it seems likely that Brunel had been lobbied by some members of the board to accept that firm's proposals. Writing to the board later, he argued that the company should put aside local interests, instead ensuring that the best engine for the job was chosen. It was essential, he concluded, to equip the *Great Western* with an engine which was 'perfection in all its detail from the moment of completion'. One of Brunel's reasons for choosing Maudslay, Son & Field was its experience in building large marine engines, and those built for the *Great Western* were large indeed. The cylinders were constructed with a 73½in diameter and a stroke of 7ft, and weighed 245 tons. These engines were originally supplied with steam from four boilers, although these were replaced in June 1844 and again in 1848. As construction continued, some changes to the original design took place, including alterations to the ship's internal layout, increasing its size by over 100 tons.

The hull was finally completed, and the ship was launched on the morning of 19 July 1837. Over 50,000 people crowded into the docks, with spectators crammed on to ships in the harbour as well as gathered on the quayside. Local newspapers reported that, when the ship started down the slipway, the assembled throng shouted 'She moves!'. Claxton broke a bottle of Madeira over the bowsprit, which was noted as being a 'demi-figure of Neptune', and the ship was named by Mrs Miles, the wife of one of the steamship company's directors, as *The Great Western*. As the large crowds made

Above: Much of the fitting-out of the *Great Western* was carried out in London, and thus, almost a month after its launch, the ship was towed down the River Avon to the Bristol Channel, for the voyage to the capital.
City of Bristol Museum & Art Gallery

their way home, 300 invited guests, including the directors of the company and prominent dignitaries, were treated to a dinner in the ship's main cabin.

It had been decided that the ship should be fitted-out in London, close to the factory of Maudslay, Son & Field, so on 18 August 1837 she was towed down the River Avon by the steam tug *Lion*, and was then accompanied for the passage around the south coast by the steam packet *Benledi*. Since the engines were not yet fitted, the *Great Western* made the passage as a sailing ship, using the four-masted schooner rig fitted at Patterson's yard. The ship's performance on the four-day trip was excellent; she handled well, and at times left the *Benledi* far behind, a good omen for the future.

With the approach of winter, the directors of the company decided to postpone the maiden voyage to the spring of 1838, avoiding the bad weather, and also allowing more time for fitting-out and testing. In London the ship was again the centre of attention, and newspapers of the day reported that visitors to the *Great Western* were amazed by 'her magnificent proportions and stupendous machinery'. By March 1838 the ship had been moved from the East India Dock, where much of the machinery had been fitted, to a berth in the River Thames. From here two trials took place, neither without

incident. In the first, on 24 March, the *Great Western* damaged another vessel after avoiding a Thames barge which had crossed her bows. Four days later she became embarrassingly stuck on a mud bank opposite Trinity Wharf, but was refloated on the tide after half an hour. These incidents aside, the ship performed well, and Brunel must have been well pleased with both the look and performance of his new creation.

Advertisements publicising the *Great Western's* maiden voyage had been published in Bristol newspapers early in March 1838, and these noted that she would set sail as soon as her engines had been fully tested, in early April. The steamship company's directors cannot, therefore, have been pleased to read advertisements for a rival service. The British & American Steam Navigation Co, whose own ship, the *British Queen*, would not be ready for the start of the new season, determined to hire another vessel, the *Sirius*, in order to claim the prize of completing the first scheduled steamship crossing of the Atlantic. As final adjustments were made to the *Great Western* at Blackwall, the 703-ton *Sirius* left London for New York, with only an intermediate stop at Cork planned. Gibbs called the ship a 'giant among cross-channel paddlers and too big for her trade', but she was not really built for the Atlantic.

Before setting out for New York, the *Great Western* had to call at Bristol to collect her passengers. The company had taken the precaution of ensuring that the ship's coal bunkers were full before leaving London, and so on the morning of 31 March 1838 she left for Bristol, pausing only at

Gravesend to allow a number of guests, including Brunel's father, to disembark. One further drama was to be played out before the ship finally began her maiden voyage. Half an hour after leaving Gravesend, a strong smell of hot oil was followed by flames and smoke appearing from around the base of the funnel. The ship was immediately run aground on a mud bank whilst Claxton, the Captain (Lt Hosken RN) and the Chief Engineer investigated.

It soon became clear that felt, which had been used to cover the boiler, had caught fire. Claxton, who had personally ensured that the ship had its own portable 'Merryweather' fire-engine, went below and at great personal risk brought the fire under control. While he was doing this he felt something heavy fall on him from above. When he had recovered from the blow, he looked down and saw the lifeless figure of a man lying face-down in the water which was now filling the engine room. Calling for a rope, Claxton picked the man up, and he was quickly hauled to safety. It was only later that Claxton discovered that the man he had rescued was his friend, Brunel. Climbing through the fore-hatch he had stepped on to a burnt ladder rung, which promptly gave way, sending him tumbling into the hold. Brunel fell nearly 20ft, striking an iron bar on the way, and, had he not struck Claxton, would certainly have been killed or seriously injured. Although Claxton may thus have saved Brunel's life inadvertently in the first instance, his prompt action in lifting the unconscious engineer out of the water almost certainly prevented him from drowning in the second. Although Brunel's injuries at first looked very serious, he was taken by boat to Canvey Island, and after a few weeks' rest he was able to resume his duties. The fire extinguished, the ship continued to Bristol, albeit with some difficulty, since a number of the stokers had jumped overboard when the fire started.

The arrival of the *Great Western* at Bristol on the evening of 2 April caused something of a sensation, since news of the fire on board had spread rapidly in the city, and all manner of rumours had been circulating as to the disaster which had befallen the ship. There was, however, little outward sign of any problem, and in the five days before she left for America, the steamship company allowed over 4,000 visitors to tour the ship. The *Great Western* was moored at Broad Pill, seven miles downstream from Bristol Docks, since her paddle wheels made her too large to pass through the dock gates of the Floating Harbour. As a result, coal, stores, cargo and other material had to be ferried out by a fleet of small boats, a slow and awkward procedure, especially when high winds and tides made operations difficult. It had been intended that the ship would leave for New York on Saturday 7 April, but bad weather delayed departure until the next morning. Even then, loading was not quite complete, and 80 tons of coal were left behind, with 600 tons in the bunkers. There were also fewer

STEAM BETWEEN

BRISTOL AND NEW YORK.
THE GREAT WESTERN,

Lieut. JAMES HOSKEN, R.N., Commander,

Is intended to sail from BRISTOL, for NEW YORK, on MONDAY, the 28th instant, on which day Steamers to convey Passengers on board will leave Cumberland Basin at Half-past Seven, A.M.

And during the remainder of the Year she is intended to Sail as follows:—

From Bristol.	From New York.
23rd March	23rd February,
18th May	20th April
6th July	13th June
24th August	1st August
19th October	21st September
	16th November.

Fares to New York, in best State Rooms, 40 Guineas; in other State Rooms, 35 Guineas.
Children and Servants Half-price.

Specie, if above £20,000, at 3/8 per Cent.; under that amount 1/2 per Cent.

Parties who have secured Freight (the whole of which is engaged), are informed that the 21st will be the latest day upon which it can be received.

Shippers of Debenture Goods will please take notice that the Certificates must be transferred to themselves, or the Goods cannot be shipped.

An experienced Surgeon is attached to the Ship.

For information, and to secure State Rooms, apply to Mr. T. M. WARD, Great Western Railway Office, Prince's-Street, London; Messrs. GIBBS, BRIGHT, and Co., Liverpool; Messrs. HAMILTON, BROTHERS, Glasgow; Mr. W. DAVIDSON, Havre; Mr. A. AUDELLE, Paris; and of CHRISTOPHER CLAXTON, at the Company's Office, Bristol.

Passengers are particularly requested to have the word "Below," in large letters, written on such portions of their Luggage as are not required in their Berths.

Freight of Measurement Goods £5 per Ton, which will be collected in New York, at the rate of 4 Dollars 80 Cents to the Pound Sterling.

No Second-Class or Steerage Passengers taken.
No Parcels can be received after the 26th instant.

N.B. Great inconvenience having resulted from Passengers bringing Merchandize on Board at the last moment as Baggage, they are respectfully informed, that in future, besides a small Portmanteau or Carpet Bag, for use on the Voyage, only 15 cube feet of *bona fide* Luggage will be allowed to each adult, and half that quantity for Children and Servants; and all Merchandize or Samples brought on board by Passengers, who have not previously engaged room for the same, will be charged at the rate of Five Shillings per foot, and if damaged, the Directors will not be responsible.

Above: One of the advertisements placed by the Great Western Steamship Company in the spring of 1838 to drum up trade for the new ship. *Author's Collection*

passengers than planned, 50 of the 57 originally booked having cancelled on hearing of the fire on the Thames.

Despite some teething problems with the engines, and rough weather which caused the one of the masts to be badly damaged, the passage to the United States was relatively trouble-free, and on the morning of 23 April the *Great Western* stopped to pick up the New York pilot, who would guide her into the harbour. Entering the port, cannons were fired, and the ship was saluted by a host of smaller boats and

Above: The occasion of the maiden voyage of the *Great Western* led to the commissioning of a number of paintings of the ship. This illustration shows the ship surrounded by a flotilla of smaller vessels. *Swindon Museum Service*

Above: The elegant if traditional lines of the *Great Western* are shown in this contemporary illustration of the vessel. At 450hp her two Maudslay engines were small in comparison with the four installed in the *Great Britain* which generated 1,000hp. *City of Bristol Museum & Art Gallery*

the *Sirius* herself, now lying at anchor. The *Sirius* had arrived the evening before, but, attempting to enter New York without a pilot, had run aground and had been forced to wait for the next tide before she could enter the harbour. In the end, after taking on more coal at the pilot station, she had only steamed in to New York 3½ hours ahead of *Great Western*; she had come very close to running out of fuel, and in the last few days her crew had resorted to burning furniture and other woodwork. The *Great Western*, however, still had well over 100 tons of coal left, averaging 30 tons per day on the voyage. Although the *Sirius* had pipped the *Great Western* at the post, it had taken her 19 days to cross the Atlantic from Cork. Brunel's steamship had made the crossing in 15 days 5 hours, and had had to travel an extra 220 miles in the process.

The crews of both ships were treated as celebrities during their stay in New York, and thousands visited the vessels. Such was the weight of numbers on the *Great Western*, that Captain Hosken had to introduce a ticket system to control the hordes wishing to see his ship. He and his crew were treated to civic banquets and other entertainments; however, the celebrations were somewhat muted following the death of the Chief Engineer, George Pearne, after he was severely scalded in an accident when inspecting the boilers.

The *Great Western* left for home on the afternoon of 7 May 1838, this time carrying a healthier complement of 68 passengers. Also on board was a cargo of cotton, bound for the Great Western cotton mills in Bristol, newspapers and mail. Fourteen days later, the *Great Western* arrived off Bristol after a passage which had enjoyed calmer weather. The crew had not, however, had the opportunity to enjoy the easier conditions, since problems with one of the crank-bearings on the port engines had caused some anxiety and a number of unscheduled stops along the way. Once again, triumphant scenes greeted the ship, and newspaper reporters were even able to complete their journey to London by Great Western train, although travel was by horse and carriage as far as Maidenhead.

Much was made, at the time and subsequently of the 'race' between the *Sirius* and the *Great Western*; however, there is little real evidence to support this idea. The two ships were not really comparable, since the *Sirius* could carry little in the way of cargo, and had only just enough coal capacity to cross the Atlantic. The *Great Western*, on the other hand, showed that a steamship could be used to cross the ocean economically with passengers and cargo, with coal to spare. Brunel had shown that the notion of transatlantic steamship travel was not the fanciful dream Dionysius Lardner had claimed, but a practical reality. In 1838 the *Great Western*, described by C. R. Gibbs as 'indisputably the first commercial ocean steamship', made five voyages to the United States, taking an average of just over 16 days to make the crossing. The return trip took an average of 13 days 4 hours. The little steamship *Sirius* made only one more trip across the Atlantic, before returning to her original duties. Other shipping companies were keen to compete with the Great Western Steamship Co, however, and the Transatlantic Steamship Co and the British & American Steam Navigation Co both ran services to New York. Neither company was particularly successful, with the latter losing its second ship, *The President*, with all hands in 1841.

The Great Western Steamship Co's prospectus had made provision for the construction of two ships, and as early as September 1838 the company announced its intention to build the second vessel, to be named the *City of New York*. It was even reported that the company had purchased a quantity of oak for the hull of this new ship, but ultimately this would not be used. Brunel and the company's directors concluded that a larger ship would offer a better return on their investment, and although original plans for the second ship showed it to be 236ft long, by November 1838 Claxton had written that the new vessel would now be 254ft in length. Some months later still, the plans were changed again, this time to accommodate the fact that Brunel's new design would now have an iron hull.

The delay in providing a companion for the *Great Western* was to cost the steamship company dear. In 1839 it lost the vitally important contract for the carriage of mail to the United States, to the new company run by Cunard. Although the Great Western company was confident of winning the tender, the fact that Cunard had four ships gave him the prospect of a reliable and continuous service, something the Bristol company could not achieve without chartering other vessels. Matters were also not helped by the fact that the *Great Western* was laid up for a three-month refit at the time of the mail contract negotiations. Further trouble was encountered where it was least expected, in Bristol; it has already been noted that the ship could not use the existing docks, as she was too large to fit through the dock gates. Having had to moor her at King's Road in the Bristol Channel, the company was somewhat taken aback to be charged £106 in harbour dues by the authorities. Furthermore, when the steamship company wrote formally to the dock company, asking that the gates be widened to

Below: A contemporary lithograph of the *Great Western* issued along with an account of her successful first transatlantic voyage. *City of Bristol Museum & Art Gallery*

THE GREAT WESTERN STEAM SHIP OF BRISTOL.

ARRIVAL
OF
The Great Western Steam Ship
IN BRISTOL.

THIS magnificent vessel, after a most prosperous voyage of 14 days from New York, anchored in Kingroad at 10 A.M. on the morning of Tuesday, the 22nd inst., having left New York on Monday, the 7th inst., 2.5 P.M.

The *Sirius*, on her passage out, was 19 days; and 17 returning. The *Great Western* (having a 20 hours' distance further to run than the *Sirius*) accomplished her outward voyage in 16 days; thus proving, beyond a doubt, her superiority, in point of speed ; and, as respects the splendor, comfort, and accommodation of the two vessels, no comparison can be instituted.

The *Great Western* is the first steam vessel that has made a DIRECT passage from a British port to the shores of America.

We are indebted for the following particulars to the *Bristol Gazette* :—

The *Sirius* brought home 17 passengers, first cabin ; 8 second cabin ; and 16 steerage : in all 51 passengers.

By the subjoined list it will be seen that the *Great Western* brought home 68 cabin passengers, and we have pleasure in inserting the following testimonials to the ability & gentlemanly demeanour of her gallant commander :—

The Great Western brought 5,553 letters and 1,760 newspapers ; also a quantity of cotton for the Great Western Factory.

Extracts from the American Papers.

Visit to the Great Western by the Ladies of New York.

Above: A handbill published by J. Williams of Broad Street, Bristol which gives details of the triumphant voyage of the *Great Western* and the alleged race with the *Sirius. City of Bristol Museum & Art Gallery*

allow larger ships to enter the Floating Harbour, it received little encouragement. The dock company's representatives replied that it would be 'manifestly unjust to be defeated of their profits by the building of ships too large to enter the harbour'. The loss of the Atlantic trade had concerned others in the city, including the Corporation, the Merchant Venturers and the Chamber of Commerce, but despite their protests, the harbour authorities remained obdurate, sealing the fate of the city docks in the longer term. The lack of a berth in Bristol also made a mockery of the original reason

Above: The mountainous seas shown in this view of the *Great Western* not only say a great deal for the strength of Brunel's designs for the ship, but also a lot about the determination and bravery of the passengers who travelled to the New World on the ship! *City of Bristol Museum & Art Gallery*

Below: Another view of the *Great Western* in heavy seas, as portrayed by an artist from the *Illustrated London News. Illustrated London News*

for building the ship, as an adjunct to the Great Western Railway. By June 1841 this had opened fully from London to Bristol, but passengers then had to travel down the river to Avonmouth before joining the ship.

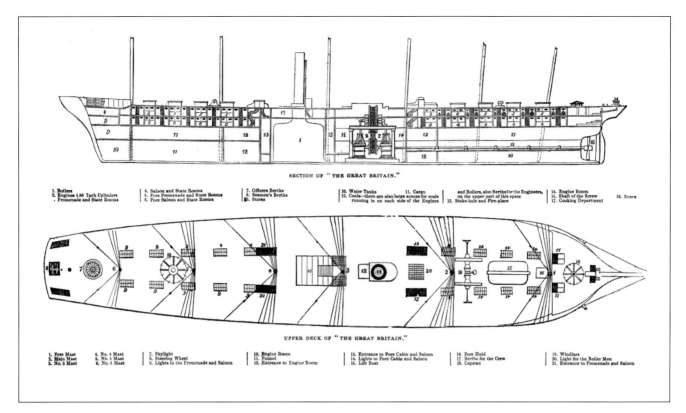

SECTION OF "THE GREAT BRITAIN."

1. Boilers
2. Engines 4.88 Inch Cylinders
. Promenade and State Rooms
4. Saloon and State Rooms
5. Fore Promenade and State Rooms
6. Fore Saloon and State Rooms
7. Officers Berths
8. Seamen's Berths
9. Stores
10. Water Tanks
12. Coals—there are also large spaces for coals running in on each side of the Engines
11. Cargo
and Boilers, also Berths for the Engineers, on the upper part of this space
13. Stoke-hole and Fire-place
14. Engine Room
15. Shaft of the Screw
17. Cooking Department
16. Screw

UPPER DECK OF "THE GREAT BRITAIN."

1. Fore Mast
2. Main Mast
3. No. 3 Mast
4. No. 4 Mast
5. No. 5 Mast
6. No. 6 Mast
7. Skylight
8. Steering Wheel
9. Lights in the Promenade and Saloon
10. Engine Room
11. Funnel
12. Entrance to Engine Room
13. Entrance to Fore Cabin and Saloon
14. Lights to Fore Cabin and Saloon
15. Life Boat
16. Fore Hold
17. Berths for the Crew
18. Capstan
19. Windlass
20. Light for the Boiler Men
21. Entrance to Promenade and Saloon

Although publicity issued by the company showed the *Great Western* to be the fastest in the trade, in 1841 she made only five voyages. The following year, in an attempt to reduce the crippling dock charges at Bristol, alternate voyages were made from Liverpool, and on 11 February 1843 the ship left Bristol for New York, the last departure of an transatlantic liner from the port for 28 years. In December 1846 the *Great Western* was taken out of service at Liverpool, having undertaken 45 voyages in eight years of service, during which 4,318 passengers were carried to the New World, and 3,357 brought back. After being laid up at Bristol for a year, the ship was sold to the Royal Mail Steam Packet Co for £24,750, and was then used for a decade on voyages to the Gulf of Mexico. In 1855, the ship was requisitioned by the government for use as a troopship during the Crimean War but, when the conflict ended, a refit was considered uneconomic and she was sold for scrap. In 1857 she was broken up at Vauxhall by Messrs Castle, the London firm of shipbreakers, and amongst those who travelled there to see the ship was Isambard Kingdom Brunel himself, by then deeply involved in the creation of a ship far larger than his first, the giant *Great Eastern*, being built not far away at Millwall.

Returning to the construction of a sister ship for the *Great Western*, it has already been mentioned that, initially, the intention of the Great Western Steamship Co was to build a new vessel of similar size. In typical fashion, Brunel was not content to take the straightforward course, and repeat his

Above:
A cross-section of the *Great Britain. Illustrated London News*

earlier design, albeit with some improvements, but determined instead to create a new and larger ship. The early history of what eventually became the *Great Britain* is somewhat complicated, but shows clearly how new ideas could be quickly adopted and incorporated into Brunel's work. The ship proposed by Brunel in November 1838 was to be around 228ft long, and had a breadth of 38ft 6in. This wooden-hulled ship would have weighed around 2,000 tons, and Brunel thus specified larger engines with cylinders of 85in to power her. There is some doubt as to whether a wooden ship of these dimensions would have been strong enough to withstand the battering of North Atlantic gales, but by January 1839 there had been a far more dramatic change, in that the hull of the new vessel was now to be made of iron. The new design was bigger again, and called for a ship 295ft long, with a breadth of 41ft, which would weigh 2,430 tons.

The decision to build an iron-hulled ship was bold, for, although an iron steamship, the *Aaron Danby*, had been constructed as early as 1821, and a number of others had been built subsequently, none was as big as the ship planned by Brunel. In 1838 the two largest iron steamers were the *Royal Sovereign*, which was 178ft long and weighed 446 tons, and the *Rainbow*, which was slightly bigger. The ship eventually named the *Great Britain* would be far bigger than either of these vessels and, as Ewan Corlett points out in his detailed history of the ship, eight times the loaded weight of her nearest rivals. It is thought that the impetus for the change in building material came partly from the visit of the *Rainbow* to Bristol in October 1838. Along with William Patterson, Claxton, who was then Managing Director of the steamship company, made a number of trips in her and other iron ships, to test their performance at sea, and to note the effect of so much metal on the working of compasses. The ship was fitted with a system of correcting magnets, which had been invented by Professor Airey, the Astronomer Royal. Writing in 1845, Claxton reported that 'iron would afford greater strength, greater buoyancy, and more capacity at less expense than wood'. He illustrated the greater capacity by arguing that, for a ship the size of the *Great Britain,* even if all the metal required to build angle-irons and ribs used in the vessel's construction were to be added to the metal plates needed for her sides, it would still be equal to an average thickness of 2½in of metal, compared with 2ft of timber.

A report was prepared for the directors, and in it Brunel, Claxton, Patterson and Guppy enthusiastically endorsed the change to iron. Further advantages were mentioned: dry rot, the 'plague of wooden ships', as one contemporary described it, could be avoided altogether, and there would be 'freedom from vermin and from the stench and unhealthy consequences of bilge water'; it was also hoped that lightning strikes, so common in wooden ships, would be avoided and that, in the event of the ship's running aground, the iron hull would prove stronger. Bearing in mind the subsequent history of the *Great Britain*, the last claim cannot be viewed without some irony. With such strong recommendation from these four men, the directors endorsed the decision to build the ship in iron, with Brunel, Claxton and Guppy acting as Building Committee, much in the same way they had with the *Great Western*. Claxton praised the directors for taking this decisive step, pointing out that, at the time they did so, they had no evidence other than that presented to them by Brunel and his team. In many ways, therefore, they too must share some of the credit for one of the most outstanding achievements of the Victorian age.

In July 1839 the keel of the new vessel, nicknamed the 'Mammoth', was laid down, and work began on its construction. Since this new ship was to be the largest paddle-steamer ever built, Brunel paid considerable attention to the design and specification of the engines. Tenders were invited for engines with cylinders of 120in diameter from

Below: Although the original negative of this rare picture has been damaged, its background clearly shows the site of the Great Western Dock where the *Great Britain* was built, albeit in later years after the demise of the Great Western Steamship Company. The dry dock is visible, as is the three-storey factory building on the left-hand side of the picture.
City of Bristol Museum & Art Gallery

Maudslay, Son & Field, which had built the engines for the *Great Western*, and from a Mr Humphreys. Brunel took some time to make recommendations to the board based on the proposals received from both companies; Humphreys' designs were substantially cheaper than Maudslay's, largely because Messrs Hall, the company which was to build the engine for him, stated that, since so much new tooling and plant would be required for the project, the steamship company itself should build the engines at Bristol. Brunel, often bold in the execution of his designs, was unusually cautious on this occasion. He had worked with Maudslay's on a number of occasions, believing it could be relied upon to deliver, and had grave doubts about Humphreys' experience. He also felt that Humphreys' prices were over-optimistic, and that the cost of building the engines would 'be fully as large or probably larger...than employing a manufacturer'.

The directors did not share Brunel's misgivings, and in June 1839 approved the plan to build and fit the engines and other equipment at Bristol. The Great Western Steamship Co then took out a 28-year lease on land next to the Floating Harbour, and in the next few years created the world's first integrated steamship works. Within the site was a purpose-built dry dock, a large three-storey factory, a sawmill and a variety of other buildings. The scale of the company's investment, at almost £50,000, should not be underestimated; the directors hoped that, once the *Great Britain* was completed, more ships would be built or repaired in the complex, allowing them to recoup their investment easily.

Once construction had begun, practical problems began to emerge, due in the most part to the enormous scale of the project; it was necessary for Humphreys to produce a far larger scale version of his patent engine. Moreover, one of the most pressing problems was that there was no hammer large enough to forge the ship's crankshaft, which was to be 30in in diameter. James Naysmyth, the engineer who was supplying much of the equipment used by the company, sketched out a drawing for a steam hammer, an invention which went on to become one of the most important of the Victorian era, and to make Naysmyth a very rich man. The dry dock itself, which was built on the advice of Brunel, was also altered, as the 'Mammoth' grew. The decision to build and fit the engines at Bristol meant that the company had to deepen the dock under the whole length of the ship as she already lay, under construction.

Construction of the ship continued apace until May 1840, when the steamship *Archimedes* visited Bristol. The arrival of a steamship in the harbour might not, at the outset, seem of great significance, but the *Archimedes* was a unique ship, one aspect of her design of special interest being the propeller. Although propellers had been used in the 18th century, it was not until May 1836 that Francis Petitt Smith properly patented the screw propeller, followed a few months later by a rival inventor, Capt John Ericsson. Both men experimented with a number of vessels, but in 1839 Smith designed and built the *Archimedes,* a 125ft steamer. After trials in the English Channel, the ship was taken on a tour of British ports, during which demonstrations were given. Claxton described the *Archimedes* as a 'handsome rakish craft', although the gearing and engines 'proved so objectionable, from the intolerable noise which it made'. Having impressed Brunel and his committee in trials in the Floating Harbour, the ship took Guppy on a trip to Liverpool and back.

Following the visit of the *Archimedes*, it was recommended that work on the engines and hull of the new ship be suspended while further experiments took place. This measure approved by the directors, Brunel arranged to have the use of the *Archimedes* for three months, in order to carry out trials with various propeller designs. The results of his work were contained in a lengthy report to the board of directors, which he personally presented in December 1840. The main conclusions of this very detailed and technical report were that the adoption of screw propulsion would result in a reduction in breadth of the ship and a substantial reduction in weight. Steering would be easier, and the simplified form of the hull would reduce sea and wind resistance and make for an altogether stronger design. Furthermore, unlike paddle wheels, the screw propeller would be unaffected by the rolling and pitching of the ship, and evidence collected by both Brunel and Claxton was presented to show that, even in very heavy seas, it was the bow of the ship which was likely to be thrown out of the water, not the stern, where the propeller was situated.

Brunel's report was unequivocal, although he was well aware of the risk he was taking, and the responsibility he bore. Writing to Guppy three years later, in August 1843, he noted that, should the project fail, his report 'would be remembered by everyone and I shall have to bear the storm'. Once again, Brunel had taken the difficult rather than the easy or familiar route; not content with building the largest iron ship yet proposed, he now introduced another new innovation, screw propulsion, and the engines necessary to operate it. With the adoption of the new arrangements, the engines proposed by Humphreys were unsuitable. Construction was well advanced, and it was reported that the hull had been completed up to the level of the paddle boxes; at this stage, Humphreys was asked to redesign the engines to suit the new screw propellers, but instead he resigned, his ambitions and plans thwarted. Shortly after, Humphreys died a broken man, his demise no doubt hastened by nervous exhaustion and his disappointment at the turn of events — matters had not been helped when Humphreys had poached a talented young engineer from Maudslay, a breach of etiquette not appreciated by Brunel. To power the *Great Britain*, Isambard then adopted the 'Triangle' engine patented by his father Marc, and he and Guppy supervised its construction in Bristol. Consisting of four inclined cylinders, the engine developed 1,600hp and drove the propeller-shaft using four large chains, which transferred power from an enormous flywheel measuring over 18ft in diameter.

With construction already underway, it was decided to drop the name 'Mammoth' and call the new ship the *Great Britain*. Her size, and the technical innovations introduced by Brunel, made her the centre of much attention very early on in her construction. Claxton wrote that: 'The sides of the *Great Britain* were scarcely visible over the walls of the yard...when naval officers, ship-builders, engineers and philosophers from all countries began to seek admittance.' In designing the *Great Britain*, Brunel took account of the need for extra strength to withstand Atlantic storms, as he had done with the *Great Western*. The bottom of the ship incorporated 10 longitudinal girders on either side of the keel, and the ship was also divided into five watertight bulkheads. These, Claxton wrote, added greatly to her strength and, if the ship were to be holed, 'her buoyancy would only be slightly affected'.

The iron plates which made up the outside of the hull were built with lapped joints, rather than being riveted together flush. Trials undertaken by the steamship company proved that this method, similar to that used on 'clinker-built' ships, was 25% stronger. The plates themselves, which measured 72in by 29in, were made at the Horsehay Ironworks at Coalbrookdale, and transported to Bristol by canal. As first built, the ship was also fitted with what was called 'Brunel's balanced rudder'. This design made steering the great ship very easy, and she was said to have handled 'like a yacht'. The original propeller was a 15ft 6in six-bladed Brunel design, but its lack of durability led to its eventual replacement with a stronger four-bladed version. Another innovation was the provision of six masts fitted with iron rigging, five of which were hinged, designed to reduce the number of men required to work the sails.

With all the changes to the designs, work on the completion of the ship was slow. Delays were also caused by financial problems; the cost of the project, again due to the many alterations, had risen sharply, and would ultimately be £125,555, double that of the *Great Western*. Bristolians, some of whom had invested in the steamship company, were less than happy with what they saw as Brunel's reckless experimentation. Thomas Latimer, an outspoken critic of the engineer in his *Annals of Bristol*, wrote that the decision not to build a straightforward sister ship for the *Great Western* was 'as imprudent as it was tardy', and that the company's decision to build the *Great Britain* gave Brunel 'full scope and leisure to indulge his passion for experiment and novelties'. There is no doubt that the success of Brunel's Great Western Railway must have played some part in encouraging investors to part with large sums of money in what must have seemed a fairly risky enterprise, even at the outset.

Despite the difficulties, Brunel's energy and enthusiasm seemed boundless, and money was found to complete the work, so that by the spring of 1843 the shareholders of the steamship company were assured that the *Great Britain* would be launched within three months, and might well be seaworthy in six. However, Isambard nearly did not witness the launch of his new ship, as on 3 April 1843 he swallowed a half-sovereign while playing with his children. The coin lodged in his windpipe, stubbornly remaining there for six weeks despite the efforts of the surgeon Sir Benjamin Brodie, who tried unsuccessfully to remove it in a tracheotomy operation. Two attempts were made to dislodge the half-sovereign using a hinged table-top, on to which Brunel was stretched and then turned upside down. On the first occasion this failed, but on 13 May, with some gentle taps to the back, the coin fell out. Writing to Christopher Claxton he remarked that: 'At 4½ I was safely delivered of my little coin — with hardly any effort it dropped out as many another has and, I hope, will drop out of my fingers.' Brunel's reputation caused the event to become national news, and regular reports were carried in the newspapers, recording his predicament.

Progress on completing the *Great Britain* was rather more prolonged; however, work had proceeded sufficiently to allow the official launch of the ship on 19 July 1843. This was a particularly significant date, being the sixth anniversary of the launching of the *Great Western*. Prince Albert was the guest of honour, having arrived from London on another of Brunel's creations, the Great Western Railway The special train had been driven by Gooch, with Brunel also on the footplate, although Gooch recorded in his diary that the journey was interrupted by 'long stops for the Prince to receive addresses'. This must have referred to a six-minute stop at Bath, where the Prince received a speech of welcome, and gave what must have been a very rapid reply! The Prince was greeted by the Lord Mayor of Bristol at Temple Meads, and was then conducted through the city amid 'many demonstrations of joy'. Arriving at the dock, the Prince inspected the ship, and the assembled guests, including Brunel's parents, then sat down to a cold banquet which was served in the patternmaker's shop. During this grand affair the dry dock was flooded and the ship was made ready to be floated out. Prince Albert did not carry out the actual naming of the vessel, instead inviting Mrs Miles, the wife of one of the directors, to repeat the task she had performed at the

Above: The launch of the *Great Britain* on 19 July 1843. Some of the thousands gathered can be seen on the hill behind the harbour as well as those who are clinging to almost every vantage point possible. *City of Bristol Museum & Art Gallery*

LAUNCH OF THE
Great Britain
GRAND STAND.
THE
BEST SIGHT

For Viewing the above, will be from the

BALLAST-WHARF, HOTWELL-ROAD,

Immediately Opposite the Head of the Ship,
WHERE THE BEST ACCOMMODATION MAY BE HAD.

For Tickets of Admission apply to Mr. SLOCOMBE, Builder, at the Wharf, or at his Yard, Cave-Street, Portland-Square.

SOMERTON, PRINTER, BRISTOL MERCURY-OFFICE. PRESENTED BY MR. JAS. SLOCOMBE, SEPT. 1911.

Above: Great Britain launch poster. *City of Bristol Museum & Art Gallery*

launch of the *Great Western*. All did not go to plan, however, and when the tow-rope from a tug to the *Great Britain* snapped, the bottle missed, and Albert completed the task instead, smashing another bottle of champagne on the bows of the ship. Many thousands of Bristolians turned out to watch the proceedings, with over 30,000 estimated to have gathered on Brandon Hill overlooking the harbour. The Prince then returned to Temple Meads almost immediately, for the return trip to London. Gooch once again drove the special train back to the capital, this time in a very rapid time of 2hr 40min, with no stops or delays. Few runs, wrote Gooch in 1892, had 'been made as quick as this, even since'.

The *Great Britain* had been fitted with masts and funnel, even though she had yet to receive her engines. Magnificent though she must have looked, a great deal more work needed to be done, and after the ceremony she was returned to the dry dock for fitting-out. During the winter of 1843 Brunel's two steamships were briefly united, while the *Great Western* was moored nearby for a refit. Progress on the *Great Britain* remained slow, however, and it was not until the end of March 1844 that the ship was finally ready to be moved into the harbour. Once fitted with her engines and other equipment, she seemed unwilling to leave her dock, and despite the water level in the dock being twice raised by 18in, she would not budge. Eventually a diver was sent down to investigate, and it was found that a piece of wood had been wedged under her hull. This removed, the ship was towed out into the harbour without incident, but in the course of this episode, it was realised that she was in fact too large to pass through the locks which linked the docks with the River Avon and the sea.

Writing in 1845, Christopher Claxton explained that it had originally been planned to move the ship through the locks without her engines; these were to be fitted elsewhere. For the 'convenience of the company...as much as on the score of economy' it was decided to reverse this decision. It seems amazing that someone of Brunel's calibre could have overlooked this rather important detail; however, he was not alone, and one can only assume that the Building Committee was so overwhelmed with all the other details of the ship's

Below: Two views of the interior of the *Great Britain*, featuring the Promenade Deck and the Saloon. *Illustrated London News*

construction that this vital consideration was not taken into account. Much debate and negotiation with the dock company then took place over alterations to the locks, necessary for the ship to pass through. Early in 1844 a deputation led by Guppy had met with the directors of the dock company, but they were insistent that a detailed specification of the work involved be produced before a decision was taken. Brunel finally produced this, and promising an indemnity for any damage that might be caused, concluded his remarks by reminding the dock company's directors that, as their consulting engineer, he would hardly recommend any course of action which would be against their interests! Matters dragged on, however, and it was not until September 1844 that a legal agreement was finally reached between the dock company and the Great Western Steamship Co.

On 26 October, the *Great Britain* was moved from the Floating Harbour to the Cumberland Basin by means of a timber cradle, which raised her high enough to pass through the inner Junction Lock. This was only the first stage, however, since she then had to pass through the outer lock into the river. As much machinery and equipment as possible was removed to lighten her, and on the morning of 11 December she was towed gently into the lock. Less than half the hull was within the lock when Claxton realised that she was scraping against the walls on either side. Fearing that she would be stuck fast when the tide began to ebb, he ordered that she be towed back into the basin. This was done with only minutes to spare, and there then followed a frantic dash

to widen the lock, in order not to miss the evening tide, which was thought to be marginally higher. Supervised by Brunel himself, an army of workmen removed the coping stones of the dock and a road bridge across the lock. This operation, which itself cost over £1,000, was a success, and the ship was quickly towed through the lock and moored on the mud outside the harbour. Writing later that night, Brunel noted that he did not want to leave Bristol 'until I see her afloat again and all clear of her difficulties'. The next morning she was towed without incident into the Bristol Channel, where she made three trial trips, on 12 December, 10 January and 20 January, each test more rigorous, to ensure that all was well.

The *Great Britain* finally sailed for London on 23 January 1845, arriving almost 40 hours later after an eventful voyage; the ship experienced some extremely rough weather, but despite this averaged 12½ knots. She was then moored at Blackwall, where she became something of an attraction, and was opened to the public. The steamship company does not seem to have been in any hurry to put the great ship into service, since she remained on the Thames for five months. The highlight of the stay was undoubtedly the visit of Queen Victoria and Prince Albert on 22 April. Escorted by Brunel and the directors, the Queen was treated to a guided tour of the ship, including the engine room, where Brunel demonstrated the workings of the engines. Like many of the

Below: The scene on the River Thames when Queen Victoria visited the *Great Britain* in April 1845. *Illustrated London News*

Above: This view of the *Great Britain* the morning after she ran aground gives little impression of the enormity of the problems facing the steamship company in removing her from the sands. *Illustrated London News*

Below: For the passengers on that fateful voyage, not only did they have to endure a miserable and frightening night when the ship was driven ashore, but they also then had the indignity of being transported to safety, along with all their belongings. *Illustrated London News*

thousands of visitors who were to flock to see the great ship, Her Majesty was reported to have been amazed by the sheer size of the vessel — at 322ft in length, far larger than any other ship then afloat.

The *Great Britain* finally left London on 12 June, this time heading for Liverpool and her first Atlantic crossing. Again, the steamship company's directors were in no particular hurry, preferring to show off the great ship at Cowes, Plymouth and Dublin, not arriving in Liverpool until early July. As previously, large crowds visited the ship during preparations for her first transatlantic passage. It was not until 26 July that she finally departed and, like the *Great Western* on her first trip, had a very small complement of passengers, carrying only 45 people. Less than 15 days later, the ship arrived in New York after a rough passage, but one in which the ship had still averaged over 9 knots. The *Great Britain* received a tumultuous welcome in the United States, the *New York Herald* describing her as a 'monster of the deep'. The question of whether Brunel's ship, with all its innovation and experiment, could actually do the job for which she had been created had been well and truly answered.

The return voyage was relatively uneventful, Liverpool being reached in a gentle 15½-day passage. Within a month the *Great Britain* was heading westwards again, this time in appalling weather, and westerly gales battered the ship on the 18-day crossing. Prior to her arrival in New York on 15 October, some poor navigation led to the ship's grounding off Nantucket, and this is likely to have been a contributory cause to damage to the propeller, which when examined in dry dock was found to be missing three blades. Repairs were effected, and the ship left New York for England on 28 October. Two days later the ship's log reported that something was once again wrong with the propeller. After reversing the engines there were 'two or three good thumps' as a blade broke off. On 6 November another blade broke off, and the ship was forced to limp home using only her sails, arriving in Liverpool on 17 November after a 20-day passage - not a bad time considering the trials and tribulations experienced by the passengers and crew.

Over the winter of 1845/6 the ship was refitted, and a more conventional four-bladed propeller installed. The *Great Britain's* third Atlantic voyage, beginning on 9 May 1846, was no less eventful, part of the after air-pump fracturing and requiring the engines to be shut down while repairs were completed. Once again, the captain had to rely on the ship's sails, which had rather fortuitously been altered and improved during the refit. The 28 passengers endured a 20-day passage, arriving in New York on 29 May. Fortunately the return trip really showed what the ship was capable of, and she made a 13-day crossing at an average speed of 13 knots. Overall, passenger numbers were not encouraging; the mishaps suffered by the ship had not helped the steamship company,

and it was not until the fourth Atlantic trip, in July 1846, that numbers increased significantly, to 110. Once more, however, the crossing was dogged by ill-luck, with faulty navigation again causing the ship to run aground, this time in Newfoundland. The return was a little better, but a potentially fast time was spoiled by one of the driving chains breaking, necessitating an 18-hour stop for repairs. Even with this delay, the passage was accomplished in only a little over 13 days.

If most of the mishaps can be put down to teething troubles, what was to befall Brunel's creation on her fifth Atlantic trip can only be described as a disaster. The increasingly positive public view of the ship had been reinforced by the record number of passengers embarking for the United States on the morning of 22 September. With 180 passengers on board, the *Great Britain* headed out into the Irish Sea, but the weather deteriorated, and at 21.00, in heavy seas, the ship ran aground in Dundrum Bay on the northeast coast of Ireland. From all the accounts of the wreck it is apparent that the hapless Captain Hosken had little idea where the ship was, thinking that she had beached on the Isle of Man. Subsequent investigations revealed that the chart he was using was inaccurate, and that his compass may have been affected by the iron hull, something which had been a worry from the earliest days of the ship's construction. Despite these extenuating factors, blame was still attached to the captain, who, as we have seen, had also allowed the ship to run off course during her previous voyage.

The passengers and crew remained on board overnight, and must have had a dreadful time; the sea continued to break over the ship for most of the night, and it was not until morning that the passengers could be disembarked. It was recorded that every horse and cart in the neighbourhood was needed to carry them and their luggage to the nearest towns. Despite the seriousness of the situation, the enormous strength of the hull had prevented the ship from breaking up and, although holed in two places and having sustained some damage to her rudder and propeller, she was still in good order. What was apparent, however, was that this situation would not last in her current position. The ship was aground on a sandy beach, but her stern and port side were directly exposed to the prevailing weather.

Claxton arrived very quickly from England, Brunel being unable to attend due to his work on other projects, particularly the South Devon Railway. A number of supposedly eminent engineers visited the ship, and submitted plans to the directors for her rescue, but, after a number of abortive attempts to refloat her, Claxton and a salvage expert, James Bremner, were forced to build a protective breakwater around the ship to preserve her against the winter weather. This was unfortunately washed away in a storm; having been rebuilt by the loyal Claxton, it was again destroyed by heavy seas. All seemed lost, and it appeared that even the Great Western Steamship Co had given up hope, its financial position being already very grave. When Brunel finally managed to travel to Ireland to view the ship, he must have been heartbroken; his creation was lying unprotected and abandoned. 'The finest ship in the world, in excellent condition...is lying like a useless saucepan,' he wrote angrily to Claxton.

Brunel channelled his anger at the incompetence and intransigence of others into a plan to save the vessel from the remaining winter storms she would likely face. His main conclusion was that the ship needed to be protected by a mass of brushwood faggots and beech spars cut from local woodland. A structure which Brunel called a 'poultice' consisted of over 5,000 faggots piled against the stern and seaward side of the ship, which were then held in place by chains, stones and iron from the ship herself. Outside, a further barrier of interwoven saplings was built to give further protection from the sea. In a formal report to the steamship company's directors in December 1846, Brunel confirmed the arrangements he had recommended to Claxton, arguing that the time was not right for discussing how the ship could be refloated; rather that her protection should be their immediate concern. Claxton was formally appointed to take charge of the salvage operation, and despite some initial problems, Brunel's protective barrier eventually held firm. Despite the harshness of his original letter to Claxton, Brunel was unstinting in his praise of him in a further report to the directors, written when the breakwater was finally finished. The work, he concluded, had 'required much skill, contrivance and unwearying perseverance'. In the spring much work had to be done before the ship could finally be refloated; hundreds of tons of sand, washed into the hold during the winter, had to be removed, and a team of shipbuilders from Portsmouth was brought in to patch the holes in her hull.

A number of attempts were made to refloat the ship on the spring tides in August 1847, but it was not until 27 August that she was finally moved off the beach and into the Irish Sea, towed by HMS *Birkenhead*. Writing to Brunel with the good news, Claxton exclaimed: 'Huzza! Huzza! You know what that means.' The *Great Britain* was found to be taking on a good deal of water, and so plans to tow her straight back to Liverpool were postponed until 29 August, to allow further repairs to take place in Belfast. When she finally left Ireland, only a combination of naval ratings and dockyard hands kept the ship afloat on her crossing to Liverpool, as the labourers taken on in Ireland to do the job were incapacitated by seasickness caused by the heavy seas encountered on the trip. On arrival in Liverpool the pumps were stopped and the ship immediately sank on to a gridiron in the Princes Dock.

A survey of the ship was carried out by Fawcett Preston, a Liverpool firm, which calculated that around £22,000 was needed to put her back into good order. The surveyors' report noted that, in similar circumstances, a wooden ship would not have been repairable, and that the iron of the frames and plates was of the 'Most excellent quality'. This

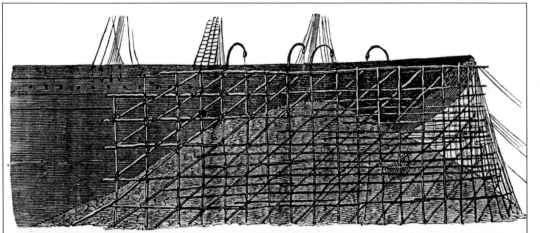

Above and left: Two views of attempts to protect the *Great Britain* from Atlantic storms. Bremner's breakwater, washed away on 9 November 1846, and Claxton's rather more elaborate and ultimately successful structure, built to Brunel's instructions. *Illustrated London News*

praise must have been cold comfort to Brunel, as the Great Western Steamship Co was in no position to repair the ship; it had recently (in April 1847) been forced to sell the *Great Western*, and now had no source of revenue, since it had been unable to attract any further shipbuilding or repair work to occupy the large Great Western Dockyard. Furthermore, the towing of the *Great Britain* from Dundrum Bay had cost £12,670 the company could ill afford and, to make matters worse, she had been underinsured, at only £17,000. Thus in April 1848 the furniture and fittings of the great ship were ignominiously sold at auction, with items such as mattresses, blankets and pillowcases all being disposed of. In September

an attempt was made to auction the *Great Britain* herself, but she failed to reach her reserve price of £40,000. It was reported the following year that the ship had been sold to the American Collins Shipping Line, but it seems that this sale fell through, for it was not until December 1851 that the *Great Britain* was finally purchased by Gibbs, Bright & Co, a shipping concern which, although based in Liverpool, had its roots in Bristol. Indeed, one of the owners, Robert Bright, had been one of those involved in discussions with Brunel in 1835, when the formation of the Great Western Steamship Co was first suggested.

Gibbs, Bright & Co paid only £18,000 for the *Great Britain*, which was a tremendous bargain, when one considers that the vessel had cost £125,555 when launched. The low purchase cost presumably allowed enough money for considerable work during her refitting, and when she made her first voyage under new management in May 1852, many changes had been made to Brunel's original designs, including new engines and a rearrangement of the passenger and cargo accommodation. Following an initial trip to New York, the ship spent the next 24 years on the route to Australia, a task for which she was well suited.

The Great Western Steamship Co, having no ships, was finally wound up in February 1852. The *Bristol Journal* reported that 'the accounts of the company show some very disastrous results'. The whole of the original share capital had been written off; the losses on the *Great Britain* were noted as £107,896, and on the works at Bristol, £42,277. Ironically, the remainder of the lease on the Great Western Dockyard was sold to William Patterson, the shipbuilder who had played such an important part in the construction of both the company's ships. It has been suggested that the directors and Brunel were cushioned from any thoughts of failure by the fact that it was predominantly shareholders' cash they were using, not their own. What is certain is that the commercial ambition shown by the steamship company, which allowed such experimentation and change to the design of the *Great Britain* over the course of her construction, whilst far-sighted in terms of progress in marine engineering, was disastrous for the investors who had placed so much faith in Brunel and the board.

Brunel had begun the venture in 1835 on the premise that the Great Western Railway could be extended to New York; the steamship company he helped form was set up largely around the network of personal contacts he had created in his years in Bristol, and both the *Great Western* and the *Great Britain* showed that fast transatlantic steamship travel was now a feasible proposition. However, overambition, poor management and some considerable misfortune combined to demonstrate that Brunel's technical innovation and design flair had simply not been enough to make the venture profitable.

Below: The stern, as work continued to recover the great ship, using hundreds of workmen to remove the tons of sand and debris which had accumulated during its incarceration. *Illustrated London News*

Right: A J. C. Bourne lithograph of a broad-gauge train. *Author's Collection*

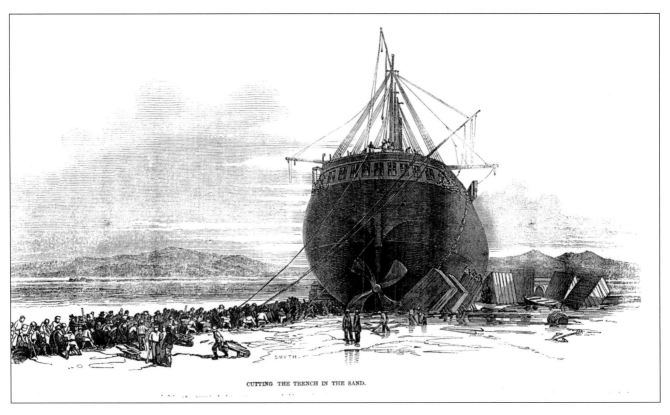

CUTTING THE TRENCH IN THE SAND.

Chapter Five

The Broad-Gauge Empire

The completion of the London-Bristol stretch of the Great Western Railway was certainly a major achievement for Brunel, but it was by no means the end of his interest in railway projects. Even during its construction, his notebooks and correspondence were full of references to work going on elsewhere on schemes such as the Bristol & Exeter Railway and the Cheltenham & Great Western Union Railway, where Brunel's broad-gauge system was spreading out into the West Country and Wales. There is some evidence that the sheer number of schemes on which Brunel was working, coupled with his involvement with the building of the *Great Britain* steamship, led to a change in the way he supervised work. Whereas during the building of the original Great Western line he had been involved in every aspect of the project, and had closely supervised everything in minute detail, after 1841 he seems to have left more to his assistants, although it is unlikely that they enjoyed too much freedom. Brunel's son diplomatically noted that 'his assistants had not perhaps so many opportunities of independent action as they might otherwise have obtained', but, he wrote, they did have the advantage of 'personal communication with their chief'.

As Brunel spent less time personally working on projects, he was to rely more on the staff he maintained at his Duke Street offices, managed after 1847 by Francis Brereton, who became Brunel's supervising engineer. Eventually, in 1848, the employment of more staff led to the purchase of the property next door, with Brunel extending his offices on the ground floor. He and his family never moved out of Duke Street, living upstairs, and much of his income was spent on furnishing and decorating the house. The grand, oak-panelled dining room was hung with paintings depicting scenes from Shakespeare, by artists such as Landseer and Brunel's brother-in-law, John Horsley. Brunel's wife Mary entertained in some style, and was said to have worn some of the most fashionable clothes of the day. With the Horsley family's existing musical connections, many of the most famous musicians and composers of the day visited the Brunels, including Mendelssohn and Chopin.

When the Great Western Railway opened fully in June 1841, the first trains to travel from the capital did not end their journey at Bristol, but continued westwards as far as Bridgwater on the Bristol & Exeter Railway. This was an independent company with its own directors, none of whom was involved with the GWR, and had been incorporated with a capital of £1.5 million in May 1836; construction of the line started the following year. In August 1840 the Great Western board reported to its shareholders that negotiations had taken place for the GWR to lease the entire Bristol & Exeter Railway for a period of five years from the opening of the completed line. The directors were confident that both parties would benefit from such an arrangement, stating that

it would speed up construction work, and enable economies to be made on locomotives and rolling stock. Short of capital, this arrangement suited the B&ER, which negotiated an annual rent of £30,000 from the Great Western, with an additional toll of a farthing a mile for each passenger or ton of goods carried.

As the Bristol & Exeter Railway's engineer, Brunel had not faced quite the difficulties he had encountered on the Great Western itself. Much of the work was supervised by William Gravatt, who had worked with Brunel and his father on the Thames Tunnel; Brunel evidently placed much trust in him, and wrote in December 1835 that, in the matter of the survey of the Exeter line, Gravatt had 'served me well'. At the Bristol end, the main engineering difficulties were a bridge over the New Cut, and an enormous cutting through Pylle Hill, to the west of the station. Although there had been plans to build a temporary station at Pylle Hill, these were abandoned, and a very simple wooden train shed was erected at right angles to Brunel's Great Western terminus. In comparison with the ornate building constructed by the GWR, the Bristol & Exeter's was rather less grand, and was known as the 'cow shed', not just because of its cheap construction, but also its proximity to the nearby cattle market! The two stations were linked by a curve, on which

Above: I. K. Brunel, portrayed by John Horsley in the 1840s, with the plans for the South Devon Railway on his desk. *National Railway Museum*

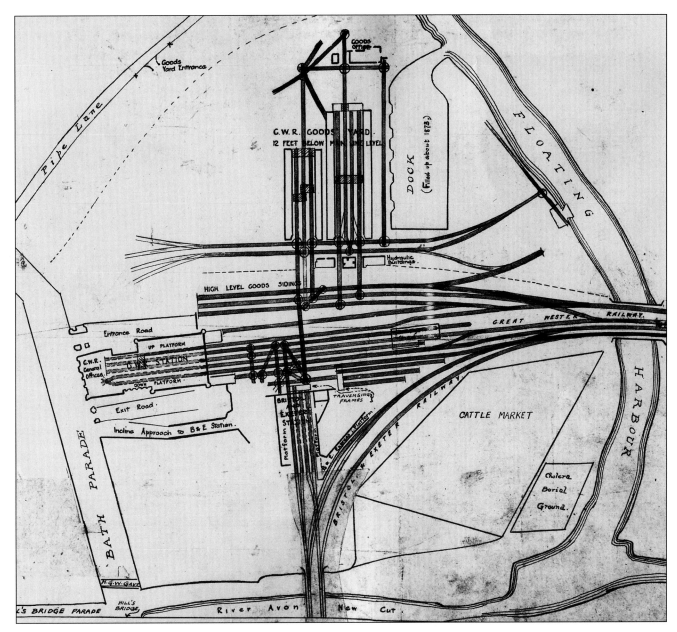

Above: Plan of the Bristol station before reconstruction, showing the cut-off line and 'express platform'. *Swindon Museum Service*

an 'express platform' was built to cater for through trains. The whole arrangement was less than satisfactory, but was not resolved until long after Brunel's death, when, in 1874, a new station was built to cater jointly for the GWR, Bristol & Exeter Railway and standard-gauge Midland Railway.

Apart from another deep cutting at Uphill, near Weston-super-Mare, the line ran on an almost level formation to Bridgwater. Construction work proceeded apace, and thus the first special train, hauled by *Fireball*, ran on 1 June 1841. Such was the excitement at Bristol that departure was delayed by nearly three-quarters of an hour, causing the locomotive to run short of water. On the next section of line, between Bridgwater and Taunton, the most notable feature was the bridge over the River Parrett, just outside of Bridgwater station. The structure designed by Brunel had a

span of 100ft, with a rise of only 12ft, a much flatter arch than his Maidenhead Bridge. This time, the design was not a success and, after movement in the foundations, he was forced to retain the wooden centring to support the structure, much to the annoyance of local people wishing to use the river. In 1843, two years after its completion, Brunel was forced to admit defeat and substitute a timber arch structure. This arch lasted until 1904, when it was replaced by a more modern girder bridge which is still in use.

Running west from Taunton, the railway burrowed through the Blackdown Hills, with a 3,280ft tunnel at Whiteball, on the border between Somerset and Devon.

Bristol. Temple Meads Station. 1870.

Construction of this section was completed by the spring of 1844, enabling the complete railway from London to Exeter to be open for traffic on 1 May. On that day, a special train was run, leaving Paddington at 07.30 and arriving in Exeter five hours later. In charge of the train was Daniel Gooch himself, and he records in his diaries that the return working left Exeter at 17.20, arriving back in London at 22.00. One of the party, the MP Sir Thomas Acland, was able to speak in the House of Commons half an hour after arriving in London, and tell the House of his rapid journey back. The experience cannot have been quite so pleasurable for Gooch, who recorded that, following his exertions on the footplate for almost 10 hours, 'next day my back ached so much I could hardly walk'. What the journey did show, however, was the increasing benefits of quick inter-city travel which Brunel's new railway system could bring.

More evidence of Brunel's interest in a broad-gauge railway which would extend far further than the original Great Western main line was the fact that, as early as the summer of 1836, he had completed a survey of a proposed railway running from Exeter to Plymouth, via Dawlish, Teignmouth, Torquay, Dartmouth, Kingsbridge and Modbury. The scheme included a number of potentially expensive features, including two large bridges, at Teignmouth and Dittisham, and was subsequently abandoned due to lack of funds. Further proposals were discussed, including a direct route

Above: A postcard view of the scene at Bristol in later years. The north end of the Great Western station is on the right, with the Bristol & Exeter building on the left. The grander building next to the B&E station contains the company offices completed in 1852. *Author's Collection*

across Dartmoor using a combination of locomotives and inclines. Lack of capital locally was still a problem in 1842, but by 1843 an agreement was brokered between the promoters of the railway and what were known as the 'Associated Companies' — the three railways which were the driving force behind the westward advance of the broad-gauge. The GWR, the Bristol & Exeter Railway and the Bristol & Gloucester Railway were persuaded to put up £400,000 capital, provided the coastal route proposed by Brunel was adopted, and that eight of the 16 directors of the new railway would be nominated by the 'Associated Companies'. This number was later increased and, although the chairman and his deputy were to be representatives of other investors, the three railways still had overall control.

The South Devon Railway Bill had few problems in the Parliamentary stage, since no rival scheme had been proposed; in July 1844 the Act was passed and the board lost little time in arranging its first shareholders' meeting, held on 28 August. The Chairman, Thomas Gill MP, a local businessman, told the shareholders that the prospects for the new company were good; traffic forecasts were healthy and, given the success of the Bristol & Exeter Railway, the amount of time and money spent to date on the choice of route would ultimately be worthwhile. He then went on to announce that the directors had decided to adopt an entirely new form of propulsion for the railway, the 'atmospheric' system. Shortly after the passing of the South Devon Act, the directors had received a letter from Samuel Clegg and Joseph Samuda, who had patented the system as early as 1838. It was claimed that atmospheric propulsion would be suited to the new line with its steep gradients, and would have advantages over the use of locomotives. The atmospheric railway would also be much cheaper, they concluded, a claim taken very seriously by the new company, which, despite the input of capital from the three broad-gauge companies, was unsure of further funding from local sources.

Gill told the meeting that Brunel, as the line's Engineer, had been consulted and was in favour of the new system. Brunel was already very familiar with the concept, and had carried out experiments using an atmospheric pipe at Wormwood Scrubs in 1840, with a view to its adoption in Box Tunnel. He had also made a number of visits to the Kingstown & Dalkey line being constructed by the Samuda Company in southern Ireland; it is thus very likely that Brunel himself had instigated the original letter to the South Devon board from Clegg and Samuda. A number of South Devon board members then visited the new Irish scheme, and were much impressed by what they saw, and more particularly by reports from the line's officials of greatly reduced operating costs.

On 19 August Brunel wrote a short report to the board, recommending the adoption of the atmospheric system, claiming that it was 'a good and economic mode'. Already worried about the steep gradients likely on the line, particularly on the Newton Abbot-Plymouth section, Brunel argued that four of the line's five inclines would have to be worked by stationary engines anyway. What he proposed was quite different from the line to which Parliamentary assent had been given; instead of a double track, he planned a single track with passing places. Since the line would not be worked by locomotives, the radius of some curves could be reduced, and gradients increased. All this, added to the saving on locomotives and their stabling facilities, would, he estimated, save the company £257,000 in initial outlay, and £8,000 per year in running costs. These figures had to be balanced by the £190,000 cost of installation, but this would be more than offset by the advantages of the atmospheric system, he argued. Apart from saving the hard-pressed company money, the new system would also be capable of running trains at much higher speeds. Brunel estimated that speeds of up to 60mph would be possible, faster than most locomotives. Another advantage was that the Samuda system was cleaner and quieter than conventional steam locomotives, which were still in their infancy, and the prospect of a journey

Above: Lord Russell, Chairman of the Great Western Railway from 1839 to 1855, and one of Brunel's strongest supporters. *Swindon Museum Service*

without dust, soot and cinders was an attractive one to many. Brunel was unequivocal in his opinion, concluding his report: 'I have no hesitation in taking upon myself the full and certain responsibility of recommending the atmospheric system on the South Devon Railway and of recommending as a consequence that the line and works be constructed for a single line only.'

There was, not unnaturally, some considerable discussion at the shareholders' meeting after this dramatic announcement. Despite the relative success of the Kingstown & Dalkey Railway, doubts were raised by at least one investor as to whether the system would be practicable over a 50-mile line like the South Devon Railway. Rather dismissively, Brunel remarked that this had already been discussed, and if he and the directors had had any doubts over its application over a greater distance, they would not be recommending it to the shareholders. After further questions, the resolution to adopt atmospheric propulsion was put to the shareholders,

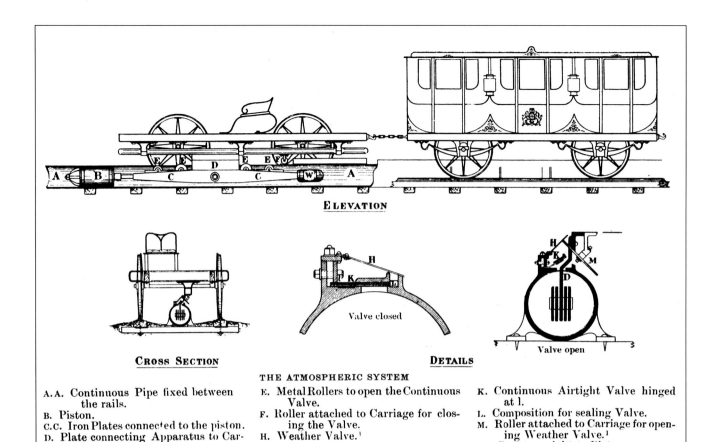

ELEVATION

CROSS SECTION

DETAILS

Valve closed

Valve open

THE ATMOSPHERIC SYSTEM

A.A. Continuous Pipe fixed between the rails.
B. Piston.
C.C. Iron Plates connected to the piston.
D. Plate connecting Apparatus to Carriage.

E. Metal Rollers to open the Continuous Valve.
F. Roller attached to Carriage for closing the Valve.
H. Weather Valve.[1]

K. Continuous Airtight Valve hinged at l.
L. Composition for sealing Valve.
M. Roller attached to Carriage for opening Weather Valve.[1]
W. Counterweight to Piston.

[1] These complications do not appear to have been in use on the South Devon Railway.

and was passed without dissent. Thus began a saga which was to end badly for Brunel, and take some of the gloss from the considerable railway work he had achieved thus far.

Despite Brunel's apparent confidence in the atmospheric system, it was not without its critics. George Stephenson was less impressed, calling it a 'great humbug'. Daniel Gooch was one of a number of prominent engineers who travelled with Brunel in September 1844 to examine the Kingstown & Dalkey Railway. Recording his visit in his diary, Gooch could not 'understand how Mr Brunel was as misled as he was', having so much faith that he could improve the system 'that he shut his eyes to the consequences of failure'. Undeterred by criticism, Brunel ploughed on, and, as the Engineer of the newly-formed Cornwall Railway as well, he endeavoured to persuade its committee to adopt both the broad-gauge and the atmospheric system, thus creating an atmospheric railway from Exeter to Penzance. This suggestion was not well received by the Great Western Railway itself, and a number of letters were exchanged between the Cornwall Railway and GWR Chairman Charles Russell, who found himself in the invidious position of having to criticise his own Engineer. The Cornwall Railway, having finally agreed to consider the

Above: An explanation of the equipment used on atmospheric railway systems. *Swindon Museum Service*

atmospheric system, argued that this would only be adopted if found to be preferable to locomotive haulage.

Before outlining the ill-fated history of the South Devon Railway, it is worth briefly describing the working of the atmospheric system which Brunel was so keen to embrace. The most notable feature of the railway was the cast-iron pipes which were laid between the rails. Services were run using 'piston carriages', which ran at the head of each train; beneath these vehicles a 15ft piston sat inside the pipe, attached to the carriage by a metal arm which passed through a 2⅛in slot in the top of the pipe. Stationary steam engines were situated at regular intervals on the route, and, when air was pumped out of the pipe in front of the train, part of the pipe behind the piston was left open, allowing atmospheric pressure to propel the piston (and thus the train along). To maintain an airtight seal on the slot, a hinged flap or valve of ox-leather was riveted to the pipe, this being weighted down with a metal strip, and an additional sealant of lime soap was also applied.

The system was particularly awkward to operate at stations, where the piston had to be removed to turn or shunt the carriage; a smaller 8in supplementary pipe was used to provide power to tow the train, using a rope attached to a piston in the smaller tube. Since the atmospheric system had no facility for reversing trains, a good deal of shunting had to be done, either by horse or manpower. At crossings and junctions there had by necessity to be breaks in the tube, each sealed by 'self-acting' valves, which allowed the piston to move from one section of pipe to another. The movement of trains was regulated by a fairly rudimentary time-interval system, whereby the pumping-engines were started when a train was due to run, and stopped after it had passed into another section. The lack of telegraphic equipment, unaccountably omitted by Brunel and the railway company, meant that the stationary engines often ran for much longer than was necessary. Braking on the train was primitive, restricted to handbrakes in the piston carriage and the guard's compartment; these were often not powerful enough to cope with the atmospheric pressure behind the piston, with the result that trains overran platforms on numerous occasions.

Progress on installing the system was slow, and in March 1846 it was reported that work on the construction of the railway itself had reached the state where the Exeter-Teignmouth section was virtually complete, except for the installation of the atmospheric equipment. In view of this situation, the directors decided to open the line in the meantime using steam locomotive power. This they duly did on 30 May 1846, when the first trains ran from Exeter to Teignmouth, hauled by two 2-2-2 locomotives which were hired from the GWR and renamed *Exe* and *Teign* especially for the occasion. The line was not opened through to Newton Abbot until December 1846.

Following the signing of an agreement with the Samuda Company in March 1845, work began on the construction of the various engine-houses a month later, with three companies, Boulton & Watt, Maudslay, Son & Field and Rennie providing the pumping-engines. Brunel had originally intended that there be three different sizes of tube, 13in for the Exeter-Newton Abbot section, 22in for the steep gradients west of Newton Abbot, and 15in for the remainder of the line to Plymouth. Over 4,400 tons of 13in pipe had been produced when Brunel, having noted problems experienced on the Croydon atmospheric line, changed his mind and opted for a 15in pipe. This decision cost the company another £31,000 it could ill afford, and costs were increased still further when it was discovered that increased pumping power would be needed to cope with the larger pipe. Inexplicably, instead of increasing the size of the engines, which were still at the design stage, he instructed the manufacturers to install auxiliary engines to achieve the power required, necessitating larger engine-houses and increasing maintenance costs.

Installation of pipework began in December 1845, but work continued slowly, much to the annoyance of South Devon shareholders. Brunel, unruffled by any criticism, told them at the end of February 1846 that the delays were in part due to his 'desire to profit by the experience of the Croydon Railway', and that he had 'postponed mere matters of detail, which has caused considerable delay in the completion of machinery necessary for our opening'. By April it was reported that only seven miles of pipe had actually been laid, although further quantities had been delivered. More importantly, the leather valve had not been fixed to the tube, allowing dirt and the effects of the weather to damage the inside of the tubes. On 16 May Brunel reassured the directors that the atmospheric system and, particularly, the valve were working well on the Croydon Railway, a claim which must be viewed with some scepticism, since four days later the whole line was closed to allow the valve to be completely replaced. The Croydon Railway did not then reopen until July and, bearing in mind his earlier claim that he intended to profit from the experience of that line, it seems unlikely that he was unaware of the problems it was encountering long before it closed. The directors were, not unnaturally, concerned when news of the Croydon closure reached them, and appear to have been further irritated by Brunel's absence from board meetings. No doubt other projects took up a good deal of his time, but his obvious discomfort and embarrassment at progress on what local people later called the 'Atmospheric Caper' meant that he must have been something of an elusive figure. On 7 July 1846 the directors formally requested that he meet with them and report on progress made and contracts placed. Some weeks later, a formal sub-committee was set up to allow Brunel and Samuda to report on progress on the South Devon Railway, and also on events on the Croydon Railway.

Work continued throughout the autumn and winter of 1846 and, despite the fact that the line was opened through to Newton Abbot, no atmospheric trains ran, even though the engines at Exeter had been tested. It was not until 26 February 1847 that the first atmospheric train ran, and, although optimistic predictions were made about the introduction of an atmospheric service to Teignmouth by the spring, work was painfully slow, delayed by problems with the pumping-engines and the valve. Matters were not helped by the news in May 1847 of the complete closure of the Croydon Railway. It was significant that Brunel was not present at a directors' meeting held a few days later, and it was left to the Company Chairman, Thomas Gill, to report that Brunel had given 'explanations of a satisfactory nature' regarding the implications of the Croydon *débâcle*. Samuda was present, however, and was able to report that engines and

Above: Four views of the atmospheric railway as portrayed by a local artist not long after it had opened. In the views shown, there appears to be little evidence of any atmospheric trains. *National Railway Museum*

tube were complete as far as Dawlish, and that test trains were operating well. Gill was again called upon to defend Brunel and the company at the next shareholders' meeting, held at the end of May, and he reported to the worried investors that tests had been successful, and that 'we have no doubts in our minds of the success of the atmospheric'.

Gill had hoped that atmospheric trains would be running to Teignmouth within a very short time of the shareholders' meeting, but it was not until August that the first experimental train was run. After this further tests were conducted, a public service being finally introduced on 17 September. A good deal of publicity was given to the introduction of the system, and it seemed on the whole to be working well. Over the next few months, services were gradually transferred from locomotive to atmospheric power, the changeover being completed on 23 February 1848.

The public euphoria surrounding the introduction of this new form of propulsion masked rather more serious problems which had emerged as the new system began operating. Difficulties were encountered in maintaining the airtight seal of both the valve and the piston-ring. These were made of leather, which proved unable to cope with the stresses of the operation and the vagaries of the weather on the south Devon coast. Neither of the sealants used on the valve was entirely satisfactory, and the leather suffered from a number of problems. It dried out in warm weather, hardened like concrete in freezing conditions, and was gnawed by rats. The resultant poor seal meant that the pumps had to work harder to make the trains run to time; increased demand and loadings on trains only compounded the problem and increased operating costs. Additional costs were incurred because, for some reason, Brunel did not install telegraphic equipment in the engine-houses; although each station was so equipped, it was left to staff to rush between station and engine-house if there was some delay or problem which necessitated an alteration to the timetable.

In the late spring of 1848, matters worsened when warm weather caused the leather valve to dry out, making it virtually impossible to maintain the seal, despite an almost continuous application of cod oil and soap; it was also found that the natural oils in the leather were being sucked out by the vacuum in the pipe. Furthermore, corrosion of the rivets holding the valve in place led to the leather tearing. Over two miles of valve had been replaced by June 1848, at a cost of over £1,000. Much of the damage was, in all probability, a result of the line's location, with the sea air accelerating the corrosion of the rivets.

On 23 May 1848 'recent irregularities in the working' caused the directors to set up a sub-committee to investigate the operation of the line, and any problems which had occurred. Although trains had been running with some regularity, the committee found serious problems with the valve, and reported that Samuda should pay for its repair, as had been agreed when contracts were first signed. Brunel was also asked to write a report giving the directors possible solutions to the problems being experienced. Once more, he was rather reluctant to act; when he showed no sign of appearing in Devon to report to the board, the directors sent a deputation to Duke Street, including Gill, his deputy Woollcombe, the GWR Chairman Russell and Buller, Chairman of the Bristol & Exeter. No formal record of the interview has survived, but an obviously embarrassed Brunel was asked to submit his written report, which he finally did on 19 August. The report was in striking contrast to the statements issued four years earlier. No improvement could be made to the existing section of the South Devon worked by the atmospheric system, he argued, without the complete replacement of the valve, and the treatment of the metal plates (attached to the leather) to reduce corrosion. Of the valve he gloomily observed: 'It is not in good working condition and I see no immediate prospect of it being rendered so.' Further work would also need to be done on the engines to increase their power, he felt, and thus the company would not be justified in spending further money, unless guarantees could be obtained from the Samuda Company that the renewed system would work well. A further humiliation was his conclusion that, in view of experience with the atmospheric system so far, he could not recommend its extension beyond Newton Abbot to Plymouth.

Samuda, having met with Gill, made an offer to improve and maintain the valve for a year, at £210 per mile; however, this proposal was met with opposition from other board members, who insisted that this arrangement should be matched by an agreement that he 'undertake the expenses of working the line'. At a board meeting held on 29 August, Gill's proposal failed to get a seconder, and the directors determined that, unless Samuda agreed to work the line at his own expense, thus removing any potential operating losses, they would suspend atmospheric working from 9 September 1848. At the subsequent shareholders' meeting, this decision led to a somewhat bumpy ride for the directors, although contemporary accounts show there was little if any opposition to the move to end the atmospheric experiment. Instead, the mood of the assembled investors was one of scorn and derision, aimed largely at Brunel and Samuda. For the workers employed by the South Devon to run the atmospheric system, the grim news from the directors was less welcome. Less than a month earlier, they had held a celebratory dinner at the Newfoundland Inn at Newton Abbot, to celebrate the opening of workshops in the town, and the success of an experimental valve, made from rubber, had led them to believe that the problems experienced were only teething troubles.

The board met again on 6 September to consider Samuda's response; this, having been deemed unsatisfactory, was the death-knell for the atmospheric. The men were given their notice, and it was decided that locomotive power would take over on Sunday 10 September. Debate over the merits of the atmospheric continued for some while after its abandonment and, although there had been a great deal of opposition to it, not all were pleased at its demise. Gill, the South Devon Chairman, resigned in November, but continued to argue the case for atmospheric propulsion until January 1849 when, at a special shareholders' meeting held in Bristol, the arguments were once again debated, in a marathon eight-hour session, and finally rejected by the majority of shareholders. All the equipment and pumping-engines had been retained until this date, but shortly afterwards arrangements were made to dispose of the pumping-engines, engine-houses and pipework, the sale eventually recovering around £81,000 for the South Devon company.

The 'Atmospheric Caper' is generally regarded as Brunel's most conspicuous failure, and undoubtedly the money lost by the South Devon shareholders makes it so. In the end, the capital cost of the scheme was estimated at £433,991, which, even when the scrap value of the equipment is taken into consideration, still left the company a legacy with which it would struggle for some years. Although Brunel might be forgiven his over-zealous enthusiasm for the introduction of a new and potentially exciting form of railway propulsion, his failure to estimate accurately the capital cost of the scheme cannot be ignored. His original estimate for equipping the whole 53-mile line from Exeter to Plymouth was only £190,000, a sum more than exceeded in the 20-mile stretch which was constructed to Samuda's system. Major misjudgements were also made by Brunel and Samuda over the power needed to run the system, which, by its very nature, could be upgraded only by the expensive replacement of pipes and pumping-engines, something which the South Devon could not afford. Brunel's absence from the project did not endear him to the directors, who resented his independence and lack of consultation with them over contracts, and the delays suffered by the scheme did little to help. Although his absence can be partly attributed to his work on other projects such as the SS *Great Britain*, there is no doubt that, by the summer of 1848, Brunel felt sufficient embarrassment and discomfort at the unravelling of the atmospheric railway to stay away from south Devon as much as he could.

Despite the failure of the atmospheric railway, the period from 1845 to 1855 was something of a golden era for Brunel. Following his exertions on the Great Western Railway and the *Great Britain,* he had achieved some considerable status within his profession, and was also reaping the benefit of his earlier hard work, earning around £15,000 per year, a

considerable sum for the time, and equivalent to around £500,000 today. As well as the larger projects described in this book, he acted as consultant on smaller projects, and gave evidence in Parliamentary Committees and court proceedings. He had also been able to secure work abroad, and travelled to Europe on a number of occasions, which did at least give the hyperactive engineer brief holidays. Mary Brunel accompanied her husband on a number of foreign trips but, complaining that sightseeing in the mountains made her dizzy, she preferred to remain at home with the children.

Having reached this point in his career, Brunel appears to have thought hard about his position and the future; still in relatively good health, he no doubt felt that things could not continue at the same hectic pace for ever. It seems highly likely that, whilst working on the South Devon Railway, Brunel had made the decision that he would eventually move out of London and settle in the West of England. In 1847 he purchased land at Watcombe, near Torquay, and for the next few years he spent what spare time he had planning both the house and the gardens with the same precision and effort usually reserved for the design of a railway. Unfortunately, the house did not progress past the foundation stage, even though both Brunel and his wife had hoped that it would be a good place for some kind of retirement (although bearing in mind Brunel's appetite for work, one wonders if this would really have been possible). Whilst work continued at Watcombe, Brunel rented a house nearby, a substantial lodge which survives today.

Two years after the purchase of the land at Watcombe, Marc Isambard Brunel died, aged 81. Brunel's parents had lived with him at Duke Street for some years, and there is some evidence, recounted by his granddaughter Lady Celia Noble, that they were not too impressed by the lavishness of arrangements at Duke Street. This is hardly surprising in the circumstances, bearing in mind the hardships they had suffered during the 'Misfortune' back in 1821. Brunel's mother Sophia continued to live at Duke Street until her death in 1855; however, it seems likely that she did not always see eye-to-eye with her daughter-in-law!

One of the other railway projects which had occupied Brunel during his work on the South Devon was another line, the Cheltenham & Great Western Union Railway. The new company held its first meeting in September 1835 and, after being appointed as Engineer, Brunel was 'asked to survey the Country, and report the most practicable line between Cheltenham and Swindon and Gloucester'. This Brunel and his assistants duly did, and within a month he was able to report that the line, which would run from St James's Square, Cheltenham, through Gloucester, and on to Swindon via the Stroud and Chalford valleys, would cost no more than £750,000. The venture did not, however, receive

Above: Purton station on the Cheltenham & Great Western Union Railway. This very early view shows that not all the structures on Brunel's lines were grand! *Purton Museum*

Parliamentary assent until the following year. Considerable opposition had come from the London & Birmingham Railway, which had planned its own line to Cheltenham from Tring, in an attempt to limit the spread of Brunel's Great Western empire. Other objections were raised by the trustees of the Thames & Severn Canal, as well as one Squire Gordon of Kemble, of whom more will be said later.

By November 1837 only minimal progress had been made, and the Cheltenham company was finding it hard to raise all the capital needed to complete the project. Some shareholders, no doubt anxious to get some return on their investment, wished to see at least part of the line completed, and so after some debate it was decided to build the Swindon-Cirencester section first, Brunel being duly instructed to place contracts. Charles Richardson, later to be the Engineer of the Severn Tunnel, was appointed as the Resident Engineer for the Swindon-Cirencester section. There is some evidence that this post was needed, since the directors on two occasions criticised Brunel for failing to produce plans and specifications for contracts, and for failing to respond to letters. As was the case on the South Devon Railway some years later, the board was forced to send a representative to see him personally, on this occasion the Chairman of the C&GWUR, Charles Sage, who questioned

whether Brunel felt able to continue as Engineer. This time, Brunel was able to convince the directors that he was committed to the project, although interestingly his diary entry for Boxing Night, 1835 reveals that he did not feel much interest in the project, and only continued with it 'because they can't do without me'.

Brunel had also noted in his diary that the Cheltenham & Great Western was 'an awkward line', and, although this no doubt referred to the heavy engineering works involved in driving the line through the Cotswolds, it may also have been a reference to the activities of Squire Robert Gordon of Kemble, who made life extremely difficult for Brunel and the company. Vehemently opposed to the railway, Gordon was one of the main objectors in the Parliamentary stage, and, since he did not wish to see any trains from his house, forced the engineer to build a 415yd tunnel to shield them from his gaze. He also insisted that no station be built within 50yd of his house; until 1872, the railway could use the station built at Kemble itself only as a junction allowing passengers to change trains for the Cirencester branch. In the intervening

Above: Mixed-gauge track, with the additional third rail added to allow standard-gauge trains to be run. The location of this photograph is not known. *Swindon Museum Service*

Below: Gloucester station in 1852. *Swindon Museum Service*

period another station was constructed, a mile up the line at Tetbury Road. The squire also had to be pacified by a payment of £7,500. The railway resisted his attempts to have the amount paid in full at the commencement of works, and the sum was paid over a number of years, with the shrewd squire managing to squeeze interest out of the hard-pressed railway as well. Brunel was less than impressed by Gordon, calling him 'cunning', but he, like the company, had little option but to put up with the landowner.

Although the capital cost of the Swindon-Cirencester section was less than that of the Cirencester-Cheltenham route, the company was still in a very weak financial state. As a result, in September 1837 they agreed to lease the line to the Great Western Railway, thus avoiding the cost of purchasing locomotives and rolling stock. Work continued on the construction of the line, and the only problems encountered on the first stretch were landslips caused by the very wet weather which had created so many problems for Brunel on the Great Western Railway. The Cheltenham & Great Western Union Railway finally opened to Cirencester on 31 May 1841, but progress on the northern section was slower. Although contracts had been placed in 1837, little appears to have been done for about two years, after which it was reported that four out of five shafts for the tunnel at Sapperton had been excavated. In the meantime, the line's directors, not content with leasing it to the Great Western, made strenuous efforts to sell it the whole operation. At first the Great Western board was unwilling to entertain anything more permanent than an extension to the lease already agreed. After what GWR Chairman Charles Russell called 'long and difficult negotiations', the company finally decided to take over the ailing Cheltenham & Great Western, perhaps fearing that, if the line were not completed, the London & Birmingham scheme mentioned earlier would be resurrected. The matter was debated at a special meeting of the Great Western shareholders on 19 January 1843, and ratified by them a week later.

With the additional capital available from the Great Western, work on the line could recommence, and the railway finally opened throughout on 12 May 1845, giving Cheltenham direct access to the capital by train for the first time. The opening of the railway to Gloucester and Cheltenham did, however, bring into sharp focus an issue which had not been far from the minds of many — the broad-gauge. During the promotion of the Cheltenham & Great Western Union Railway Bill in Parliament, the standard-gauge Birmingham & Gloucester Railway was also proposed and, both companies wishing to run trains between Gloucester and Cheltenham, they agreed to share a line. This would eventually be the first example of what became known as 'mixed-gauge' track, with both standard- and broad-gauge trains sharing track laid with an additional (third) rail.

The operational realities of running trains on the mixed-gauge may have been complicated; however, it was at Gloucester, where the first real 'break of gauge' occurred, that the first skirmishes in what became known as 'The Battle of the Gauges' took place. The first signs of trouble had been seen in the activities of the Bristol & Gloucester Railway, a line which until 1842 was intended to be a standard-gauge affair. However, under some pressure from Brunel, who conveniently was also its Engineer, the company decided in April 1843 to change its plans and adopt the broad-gauge. This change enabled the company to use the Great Western's station at Bristol, and also the Cheltenham & Great Western Union Railway's broad-gauge line already planned from Standish Junction to Gloucester, but it must also have been a reaction to the possibility of having Great Western opposition at either end of its line. Brunel had also persuaded the Bristol & Gloucester company that the break of gauge at Gloucester, between the broad-gauge and the standard-gauge of the Birmingham & Gloucester Railway, could be easily overcome by a simple arrangement to transfer goods from one wagon to another. Passengers would, he argued, 'merely step from one carriage into the other and on the same platform'.

As usual, Brunel's rather offhand remark was not completely satisfactory, and the simple arrangement he mentioned for goods handling, although sketched in his notebooks, never appeared. Delays and disruption occurred in the small goods depot, and the problems experienced were subsequently exaggerated and exploited by Brunel's opponents. Further difficulties arose when the Bristol & Gloucester joined forces with the Birmingham & Gloucester Railway in January 1845. The new Bristol & Birmingham Railway then set about negotiations with the GWR with a view to extending the broad-gauge to Birmingham. No satisfactory deal could be agreed, and the day after GWR Company Secretary Charles Saunders had told the Bristol & Birmingham that his offer was final, the line was purchased by the Midland Railway, thwarting Brunel's aim of broad-gauge travel from Bristol to Birmingham, and confirming Gloucester as a break of gauge. McDermot records that Brunel's dream was lost for a paltry £5 per share; however, the financial ramifications were the least of Brunel's worries, with the whole future of his broad-gauge railway cast in doubt.

Two further battles would be fought before the gauge question could finally be settled. By 1844 two more new lines were proposed in the area north of Oxford, which itself had been reached by railway only that same year. Both lines were of significance, to Brunel and to his opponents, since they threatened to carry the broad-gauge deep into the heart of what its supporters disparagingly called 'narrow-gauge' territory. The first of these was the Oxford, Worcester &

Above: The break of gauge at Gloucester as seen by the *Illustrated London News.* No doubt some artistic licence was used to argue the case against Brunel's broad-gauge. *Illustrated London News*

Above: If conditions for passengers were bad, then the transshipment of goods was an even greater problem, with delays and damage to goods common. *Illustrated London News*

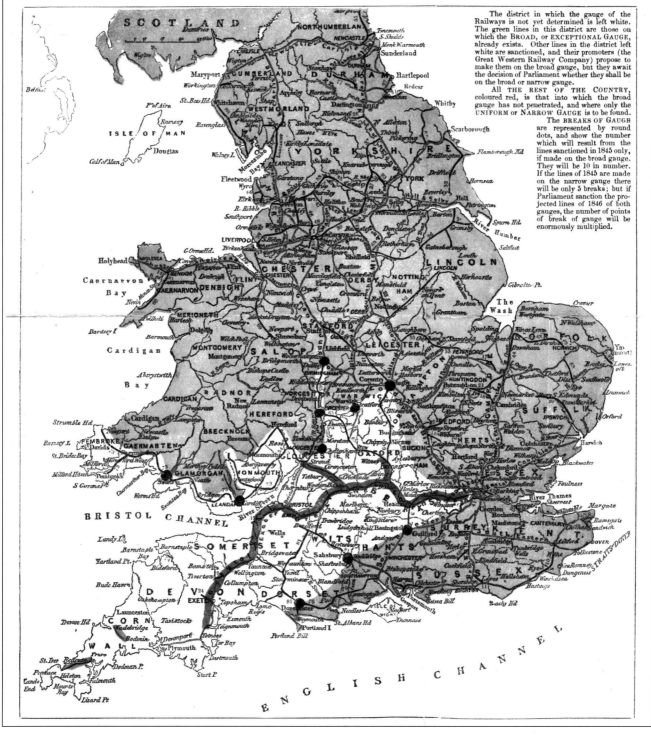

MAP OF THE DISTRICTS
OCCUPIED BY BROAD AND NARROW GAUGE RAILWAYS RESPECTIVELY,
SHOWING ALSO THE BREAKS OF GAUGE,
Where Passengers and Goods must be shifted from one Carriage to another.

The district in which the gauge of the Railways is not yet determined is left white. The green lines in this district are those on which the BROAD, or EXCEPTIONAL GAUGE, already exists. Other lines in the district left white are sanctioned, and their promoters (the Great Western Railway Company) propose to make them on the broad gauge, but they await the decision of Parliament whether they shall be on the broad or narrow gauge.

All THE REST OF THE COUNTRY, coloured red, is that into which the broad gauge has not penetrated, and where only the UNIFORM or NARROW GAUGE is to be found.

The BREAKS OF GAUGE are represented by round dots, and show the number which will result from the lines sanctioned in 1845 only, if made on the broad gauge. They will be 10 in number. If the lines of 1845 are made on the narrow gauge there will be only 5 breaks; but if Parliament sanction the projected lines of 1846 of both gauges, the number of points of break of gauge will be enormously multiplied.

Above: Map of the broad and narrow (standard)-gauge lines. *Author's Collection*

NEW RAILWAY
FROM CHELTENHAM TO OXFORD.

To the Owners and Occupiers of Property along the Line.

BEWARE OF THE GREAT WESTERN!

THEIR AGENTS

ARE NOW ACTIVELY EMPLOYED

In Calling on the Owners and Occupiers to induce them to join the
Great Western

In Opposing the Line to Oxford.

I warn you against their insidious designs.

Give them no Information or Assistance ;

AND

CAREFULLY AVOID SIGNING ANY PAPER

Presented to you under any pretence whatever.

W. H. GWINNETT.

CHELTENHAM, 27th December, 1852.

Above: Not all railway schemes were welcome, as this handbill shows! *Swindon Museum Service*

Wolverhampton Railway, its committee of management including three Great Western directors, one of whom was the redoubtable Thomas Guppy, of whom we have already heard much. Although its promoters hoped to generate much income from passenger traffic, the prospect of linking the rich, industrial West Midlands with the capital was more attractive, and the prospectus, issued in September 1844, noted that the new line would serve not only collieries, ironworks and other industrial concerns, but also a large agricultural region. There was no region 'now unoccupied by Railway which affords so great a prospect of remuneration', the prospectus boasted. Brunel, having surveyed the route, estimated the cost of the venture to be around £1,000,000. The map included in the prospectus also showed the route of another line linking Oxford and Rugby by way of Banbury, this being the Oxford & Rugby Railway.

Although the promotion of these broad-gauge railways was challenged by Brunel's opponents on the grounds of the gauge question itself, much of the opposition was based on a far more straightforward argument, that of commercial profit. Both lines were planned to run deep into the heart of territory jealously guarded by the Great Western's rivals, which were unwilling to let it take advantage of the potential income and traffic such new lines might generate. Not surprisingly, the London & Birmingham Railway, which stood to lose most, was a vociferous opponent of the new railways, and promoted its own rival scheme, the London, Worcester & South Staffordshire Railway. The large number of railways being planned during this period meant that both lines promoted by Brunel and the Great Western were considered first by a panel of five Commissioners of the Board of Trade, known at the time as 'The Five Kings', whose task it was to weed out some of the more speculative ventures then promoted. This body's initial findings, issued in January 1845, came out firmly against the broad-gauge, but undeterred, the promoters put Bills for both lines before Parliament, and the same committee was again asked to consider the plans, along with those of the GWR's standard-gauge rivals, in May. It was recorded that, over a period of 19 days, 102 witnesses, including Brunel, Stephenson, George Hudson and Daniel Gooch, gave evidence.

This time the committee ruled in favour of the broad-gauge schemes, but the victory was somewhat short-lived. When the committee's report was debated in Parliament some weeks later, Richard Cobden MP, the redoubtable advocate of free trade and opponent of the broad-gauge, persuaded the House that the whole question of a uniform gauge for railways should be investigated by a Royal Commission. Sensibly, the government engaged three Commissioners who could rightly be said to have had no particular view on the matter, and could be expected to come to a fair decision: Sir Frederick Smith, the first Inspector-General of Railways, Peter Barlow, Professor of Mathematics at The Military Academy at Woolwich, and George Airey, the Astronomer Royal. The Commission was asked to recommend whether future railway Acts should contain reference to a uniform track-gauge, to consider the practicability of creating a uniform gauge for the country as a whole, and to suggest any solutions it might have for the 'break of gauge' problem.

The Commission reached most of its conclusions after interviewing expert witnesses, the list of whom reads like a railway *Who's Who* of the period. Not surprisingly, hardened opponents of the broad-gauge, such as Robert Stephenson, missed no opportunity to criticise it. Stephenson argued that 'its introduction has involved the country in very great inconvenience', with the break of gauge being its greatest disadvantage, as well as the increased construction cost of broad-gauge lines and rolling stock. Brunel was called to give evidence on Saturday 25 October 1845, answering a total of 200 questions. After reminding the Commissioners why he had chosen the broad-gauge in the first place, he was asked if, having designed the Great Western, he were to do so again, he would still choose the 7ft gauge. In answering that question, Brunel replied, he was likely to be accused of recklessness, since 'I should rather be above than under 7ft now, if I had to reconstruct the lines'. To arguments that broad-gauge railways were proportionately larger and therefore more expensive to build, Brunel claimed that differences in cost between broad and narrow were negligible.

On the question of a uniform gauge for the whole country, Brunel was less clear, contradicting himself in the course of one answer by arguing that there would be some advantage to a 'similarity of gauge', since it would rid railways of the problem of changing from one system to the other, but that a great deal of progress had been made on railways through competition between promoters of the rival gauges. Adrian Vaughan has rightly argued that it was the quality of Gooch's locomotives and the well-engineered line designed by Brunel which brought about improvement on the Great Western, not the track-gauge, so Brunel's assertion that competition brought about improvement was not entirely satisfactory. In relation to the much-discussed problem of the break of-gauge, and the inconvenience and expense it caused, it was clear from his replies that Brunel did not see the Great Western and its associated railways as part of a national railway network — rather a regional system within which passengers and goods could travel, changing at particular locations to other lines. 'If a network of railways...over England is made, I think it will be impossible that passenger carriages can be running in all directions over that network without changing,' he argued, returning to the theme that: 'The spirit of emulation and competition will do more good

Above: Gooch's pioneering 'Firefly' class locomotive. Based heavily on the best features of the Stephenson 'Patentee' locomotive, these powerful and reliable engines were the mainstay of the Great Western locomotive fleet in the early days of the line and were used in the infamous Gauge Commission trials. *National Railway Museum*

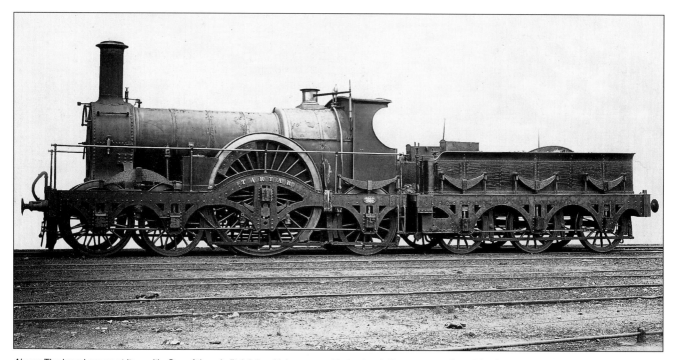

Above: The broad-gauge at its zenith. One of the rebuilt 4-2-2s which were used to haul main line expresses from Bristol to the West Country until the abolition of the broad-gauge in 1892. *Swindon Museum Service*

Great Western Railway
LOCOMOTIVE DEPARTMENT.

CONVERSION OF GAUGE
BETWEEN THINGLEY JUNCTION & DORCHESTER,
Including the Salisbury, Bathampton, Radstock, Wells, and Bridport Branches.

NOTICE TO ENGINEMEN AND FIREMEN.

Every Engineman working over the above-mentioned Lines, during the Conversion of the Gauge, will be supplied with a copy of the printed TIME TABLE AND GENERAL INSTRUCTIONS issued for the use of the Company's Servants, for which his signature will be taken. No Engineman must on any account take charge of an Engine or Train on any part of these Lines or Branches after Monday, 15th June, or subsequently during the Conversion, who has not previously received and signed for, the Time Table and General Instructions.

The Enginemen are requested to make themselves thoroughly acquainted with the Time Bill and General Instructions, and they will be expected to read them through with great care as soon as they receive them, and if they meet with anything which they do not properly understand, or which they think requires explanation, they must at once apply to their Superintendent, or Foreman, or one of the Locomotive Inspectors on duty, in order that there may be no misunderstanding whatever on the part of the men engaged in this work.

The particular attention of the Enginemen is directed to clause 15 of the General Instructions on Page 12 as to stopping " *dead* " at the end of each Section ; also to Clause 47, Page 15. ALL FACING POINTS MUST BE APPROACHED WITH VERY GREAT CAUTION, and at the Crossing Places the greatest care must be taken to have the Engine or Train so under control, as to prevent the possibility of overshooting the Points.

Enginemen and Firemen must keep a constant look out for any signal that may be given by the Permanent Way Men or others, whether by means of a Red Flag or other Hand-signal : They are to proceed with great caution, especially on approaching curves, when they must take care to sound their small whistles, so as to give timely warning of their approach to the men working on the Line.

Enginemen and Firemen are particularly requested to be with their Engines in good time, and to bestow the greatest care upon the oiling and examination of the working parts, Axle Boxes, &c. They must be particularly careful to have a good supply of coal before starting from each end, and to fill up their tanks at every watering place, so as to be fully prepared for any unexpected stoppage or delay. So much depends upon the Engines being in the best possible working order, that it is hoped very great attention will be paid to this matter, MORE ESPECIALLY WITH THE NEW NARROW GAUGE ENGINES.

Firemen are also required to make themselves well acquainted with the Time Table and General Instructions, and the Enginemen must afford them every opportunity of doing so. As far as possible the Enginemen and Firemen should read the instructions TOGETHER, so as to obtain a perfect knowledge of them before commencing to work the single line.

J. ARMSTRONG.

Engineer's Office, Swindon.
13th June, 1874.

Above: A handbill issued by the Great Western Railway to give locomotive staff details of the gauge conversion of much of the route south of Bath in 1874.
Author's Collection

Above: The scene at Swindon after the end of the broad-gauge. As well as a large locomotive fleet which had to either be converted to standard-gauge or scrapped, there were also thousands of carriages and wagons which had to be dealt with at Swindon, and extra land was purchased to accommodate them. *National Railway Museum*

Below: Converting the broad-gauge near Plymouth, May 1892. *National Railway Museum*

than that uniformity of system which has been so much talked of the last two or three years.'

Although passengers were much inconvenienced by having to change trains at places like Gloucester, the impact on goods traffic had more serious implications for industry. The average delay for goods at Gloucester was estimated at between 4½ and 5½ hours, during which time goods were manhandled by railway staff from one train to another, with much material either lost or damaged in the process. Brunel admitted that 'some inconvenience may occur', but argued that he had a solution for the problem which involved the use of what we might today call container traffic, whereby wagon bodies could be lifted from one wagon chassis to another at transshipment depots. If this method were not suitable, Brunel also suggested the use of a broad-gauge transporter truck, with narrow-gauge rails set into it, to carry, as he put it 'waggon and all'. Asked if this idea would cause problems, he dismissed the questioner, stating that there would be little practical difficulty. What he failed to mention was that, although the broad-gauge was built to generous proportions, its bridges were little bigger than their standard-gauge counterparts, so it was unlikely that this combination would fit under them.

Before completing his session as a witness, Brunel had to defend himself against some awkward questions regarding other railway projects in which he was involved, where he had not recommended the use of the broad-gauge. As well as the railway from Genoa to Turin, he had also, in 1836, become the Engineer of the Taff Vale Railway in South Wales, which had been built as standard-gauge from its inception. To critics who pointed out this inconsistency, Brunel merely argued that the main advantage of the broad-gauge was that its trains could run at consistently higher speeds, and had rolling stock which was by its nature, of greater capacity and more comfortable. On both the lines mentioned, high speeds were not a high priority, so, he concluded, they did not necessarily have to be broad-gauge.

Before the end of proceedings, Brunel managed to persuade the Commissioners to arrange a series of trials to test the worth of locomotives of both gauges. A Gooch 'Firefly' class, *Ixion*, was matched against a Stephenson 'Long Boiler' locomotive, No 54. The broad-gauge engine outperformed its Stephenson rival easily in terms of both average and top speed, with the unfortunate No 54 even derailing itself on one trip. The data obtained on these trials was added to a wealth of other statistical information supplied by both sides; costs of construction, running costs, locomotive and train performance and traffic statistics were all submitted, and the Commissioners had the unenviable task of absorbing and considering all the conflicting material, as well as the testimony of the various witnesses called.

In 1846 the Commissioners finally reported. They prefaced their findings by noting that they could find little to fault with the broad-gauge itself; there was little doubt that, with regard to safety, speed and 'the convenience of passengers', it was superior to the standard-gauge, and that 'the public are mainly indebted for the present rate of speed and the increased accommodation of the railway carriages to the genius of Mr Brunel and the liberality of the Great Western Railway'. Then came the bad news. The impressive performance of broad-gauge express trains was inescapable; however, they were for the accommodation of 'a comparatively small number of persons'. This was of less importance than the 'general commercial traffic of the country', and therefore, in the matter of the transportation of goods, the Commissioners believed that the standard-gauge was better suited for this purpose. There was little doubt that, amongst the welter of statistical information that had been submitted to the Commission, one of the most important was the fact that, up to July 1845, only 274 miles of broad-gauge line had been built, compared to 1,901 miles of standard-gauge, and given this statistic, the question of a uniform gauge was already a pressing reality. Thus, the Commissioners ruled: 'If it were imperative to produce uniformity, we should recommend that uniformity be produced by an alteration of the broad to narrow-gauge.'

Brunel's son reports that the decision caused dismay amongst supporters of the broad-gauge, who had hoped and expected that they would win the day. Incensed, Brunel, aided by Saunders and Gooch, quickly wrote a 43-page pamphlet *Observations on the Report of the Gauge Commissioners*. In it, Brunel refuted many of the conclusions reached, and accused the Commissioners of ignoring the evidence which the supporters of the broad-gauge had submitted. 'Facts are stubborn things', he wrote, and in the course of the pamphlet, which was sent to every Member of Parliament, he carefully exposed every mistake and assumption made. Repeating the arguments already discussed in the inquiry itself, Brunel argued that it would be wrong for the gauge of lines already sanctioned by Parliament to be now changed, and there should be 'a strong protest against any legislative interference with the broad-gauge system'. Other pamphlets and tracts were published, many strongly opposed to Brunel and the broad-gauge. One such publication was *The Broad Gauge, the Bane of the Great Western Railway Company*, written by '£.s.d.'. In it the anonymous writer argued that, following the decision of the Gauge Commission, the 'Broad Gauge partisans' should 'retire amid a flourish of trumpets'. No-one would suffer loss of reputation, the tract continued, least of all Brunel, who, in the construction of the Great Western Railway, had displayed abilities 'of no common order', and had established a reputation 'resting on a more solid basis than so many inches of gauge'.

Left: The cartoon which was printed in *Punch* magazine in 1892, when the broad-gauge was finally abolished, with the ghost of Brunel hovering over its grave. *Swindon Museum Service*

Above: The end of an era: an evocative photograph of the Great Western 'Single' *Dragon* taken on 20 May 1892. *Author's Collection*

Brunel's abilities as an orator and lobbyist for his cause have already been demonstrated on a number of occasions in this volume, and there is some evidence that his efforts may have caused Parliament to water down some of the conclusions of the Royal Commission, with the passing of 'an Act for Regulating the Gauge of Railways' in July 1846. Although 4ft 8½in was specified as the gauge of all new railways in Great Britain, the new Act further specified that this gauge would be the standard, unless any present or future Act had a 'special enactment defining the gauge' contained within it. This clause left the door open for the building of further broad-gauge lines, a process which was to continue for some years. The two lines which had precipitated the debate in the first place, the Oxford, Worcester & Wolverhampton and the Oxford & Rugby, were allowed to proceed as planned.

The subsequent history of the OW&W was far from straightforward, and not withstanding severe financial problems encountered by the company, Brunel himself became embroiled in a far more dramatic situation in 1851 at Mickleton, near Chipping Campden where a tunnel was being driven under the Cotswolds. A dispute over remuneration for the contractor engaged to build the tunnel, Robert Mudge-Marchant, escalated into a full-blown confrontation. On the evening of 17 July 1851, Brunel arrived at the site with hundreds of navvies, determined to evict Rudge-Marchant in lieu of the money it was claimed he owed. The local magistrates had been alerted and the great engineer was warned that he was causing a breach of the peace. The stand-off between the two groups was defused only by the reading of the Riot Act, and Brunel retreated, returning a day later with reinforcements. With over 3,000 navvies spoiling for a fight, it was remarkable that only minor skirmishes took place and that there were no fatalities. With such a show of strength, Rudge-Marchant was faced with little option but to withdraw, on the promise of an arbitrated settlement by Brunel. The whole episode does show a very different, cavalier side of the engineer, and is all the more amazing when one considers that he had enrolled as a Special Constable in Bristol to prevent disorder and unrest some 20 years previously!

It seemed that Brunel and the Great Western had won a partial reprieve for the broad-gauge, but, if anything, the

Above: Chepstow station on the South Wales Railway. Some artistic licence is revealed in the lithograph, particularly since the locomotive and carriages appear to have wheels on only one side! *Swindon Museum Service*

Gauge War defined the limits of Brunel's empire; any thought of its penetrating as far north as the Mersey was now gone. Furthermore, as the national railway network grew ever larger, the Great Western found it harder and harder to maintain the haughty isolation it had once enjoyed. Mixed-gauge track was increasingly introduced and, with the acquisition of standard-gauge lines by the Great Western in the 1850s, the long-term future of the broad-gauge was in doubt. From the 1860s, a programme of conversion began, but the broad-gauge was not completely eradicated until 1892, some 33 years after the death of its creator. The end of Brunel's broad-gauge brought a series of articles and eulogies in the press, including the following anonymous poem:

Lightly they'll talk of him now he is gone
For the cheap Narrow Gauge has outstayed him;
Yet Bull might have found, had he let it go on,
That Brunel's big idea might have paid him.

But the battle is ended, our task is done —
After forty years fight he's retiring
This hour sees thee triumph O' Stephenson!
Old broad gauge no more will need firing.

One railway which emerged unscathed from the whole Gauge War episode was the South Wales Railway, to which

Brunel had been appointed Engineer in 1844. The line was part of the through line from London to South Wales which he had planned some 10 years before, and was built with much support from Irish interests which saw the proposed line, terminating at Fishguard in West Wales, as an opportunity to generate much Anglo-Irish business. In the event, Fishguard was not developed by the Great Western until the 20th century, the potato famine in Ireland causing earlier plans for its construction to be abandoned. It was thus decided that ferry services from Waterford should call at Milford Haven, this becoming the western terminus of the new railway which ran east for over 200 miles, linking Carmarthen, Swansea, Cardiff and Newport, and turning north to skirt the banks of the Severn Estuary. Brunel had originally hoped to avoid Gloucester altogether, planning to cross the river by a bridge at Hock Cliff near Awre. Difficulties with the Admiralty finally led to the abandonment of this idea, and Brunel had no choice but to route the line through Gloucester. Space does not permit a full description of the railway, which as designed by Brunel contained numerous bridges, including a number of timber structures, the origin and design of which will be described in the next chapter. It did, however, include the bridge at Chepstow, a feature which merits more detailed analysis, containing as it did unique features which were to be developed by Brunel in later years, at Saltash.

Brunel's bridge at Chepstow was 600ft long. Approaching the River Wye from the Newport side, the railway was carried across the low, muddy bank of the river on three 100ft spans, each supported by a pair of 6ft-diameter cast-iron cylinders. To support the bridge, these cylinders were sunk to a depth of around 50ft through the mud, gravel and clay, until solid rock was found. The tubes which made up the cylinders were sunk by weights placed on top of them, which forced them downwards while workmen excavated inside. This process was a hazardous one, and the men working inside the tubes had not only to contend with floods of water, which were made worse by the extremely large tidal range on the River Wye, but also with the influx of fine, soft river-silt. Problems were also experienced when the tubes came into contact with large boulders of conglomerate rock, one of which caused one of the cylinders to crack. Before work could be completed, a wrought-iron hoop was made and fixed to the damaged tube to effect a repair. Once the tubes were completed, they were filled with concrete and gravel, making an extremely strong base for the bridge.

At the end of the three piers which passed over land, the main bridge structure was supported on a further six cylinders, arranged in two rows of three. On top of this was a 50ft tower constructed of cast-iron. At the other end of the main bridge-span, another tower, this time built of masonry, was built on the edge of a 100ft-high limestone cliff, which formed the bank of the River Wye on the Gloucestershire side. The bridge built between these two sets of towers was an ingenious design, which not only coped with the rather unusual nature of the landscape at this point, but also ensured that the 50ft clearance required to allow ships' masts to clear the bridge at high tide was adhered to. Between the towers were fixed two 312ft-long wrought-iron tubes, from which two separate bridges carrying the tracks were hung by suspension chains. Vertical struts were added, not only to support the tubes, but also to give additional rigidity to the whole structure. The 9ft-diameter tubes weighed only 138 tons each, and were of cellular construction; this was not the first time such tubes had been used by Brunel for bridge construction, a rather smaller-scale example having been

Below: Brunel's bridge at Chepstow. The awkward location, with cliffs on one side and low riverbank on the other, can clearly be seen. *National Railway Museum*

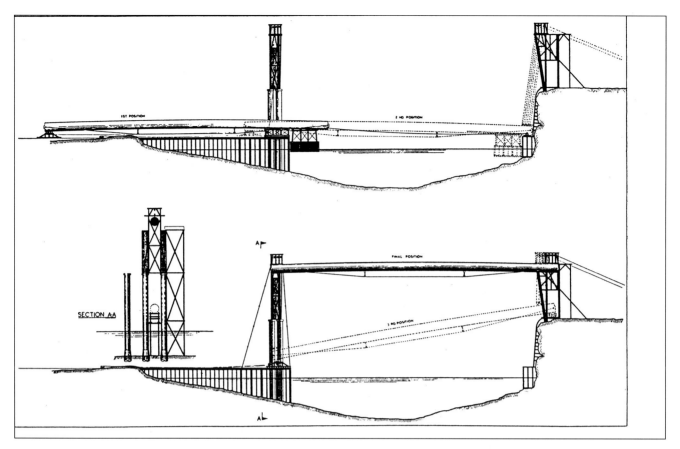

Above: A diagram showing the four stages of moving and raising the tubes of the Chepstow Bridge into position. *Swindon Museum Service*

built for a bridge across the dock gates in Bristol some years previously. Since the wrought-iron roadway girders containing the tracks were wider than the tubes themselves, the suspension chains were splayed outwards, and Brunel was thus forced to add further cross-bracing and strips of gunmetal where friction between chains and struts occurred. The enormous 20ft-long suspension chain-links were manufactured at the Seville Iron Works in Dublin, and were each forged in one piece. In typical fashion, Brunel travelled to Dublin to show the men making the links exactly how it could be done.

The bridge-tubes were built parallel to the Wye on the riverbank, and, when completed, were manoeuvred until they were at right angles to the river. A railway track was constructed on a wooden platform, and the bridge was laid on this staging, supported by railway trolleys. The operation to raise the first tube, planned in meticulous detail by Brunel and his assistant R. P. Brereton, began on 8 April 1852, when six pontoons were positioned in the river at the end of the tube. As the tide began to rise, the bridge-tube was lifted up by the barges. The tube was then towed across the channel and, once lifting-chains had been attached to both ends, the great structure was hauled up to rail level. Within a day it had been placed in position on top of the tower itself. The bridge was opened for traffic using the completed span on 14 July, and by August 1852 the *Illustrated London News*

reported that the second tube was complete and the process could be repeated. It was not until 18 April 1853 that the whole structure was finished and both sides of the bridge were open to rail traffic. The new bridge was, as might be expected, a bold and original solution to a particular engineering problem, but, unlike some of Brunel's work, this bridge was also an economical solution, costing only £77,000.

The impact of the fully-opened South Wales Railway was dramatic: in 1850 the journey from Swansea to London took over 15 hours, using a combination of coach, ferry and railway. With the opening of the South Wales line, this journey could be accomplished in little more than five hours, with rather more comfort than before. The construction of the line west of Swansea was not fully completed until the spring of 1856, with the section between Haverfordwest and Milford Haven the last to be built. Unlike Brunel's bridge at Saltash, the Chepstow structure has not survived. The passage of trains across the suspension bridge took its toll, and after around 90 years of use, the structure began to show signs of fatigue. Although some repairs were carried out after World War 2, the bridge was completely rebuilt in 1962.

Above: Chepstow Bridge from the Welsh side. The construction of the supporting piers and the bridge itself can be clearly seen. *Swindon Museum Service*

Below: Brunel's bridge at Chepstow was not his first use of wrought-iron tubes. This girder bridge originally spanned the entrance to the Floating Harbour at Bristol, but now lies out of use next to the North Lock, below the modern Cumberland Basin bridge. *Author's Collection*

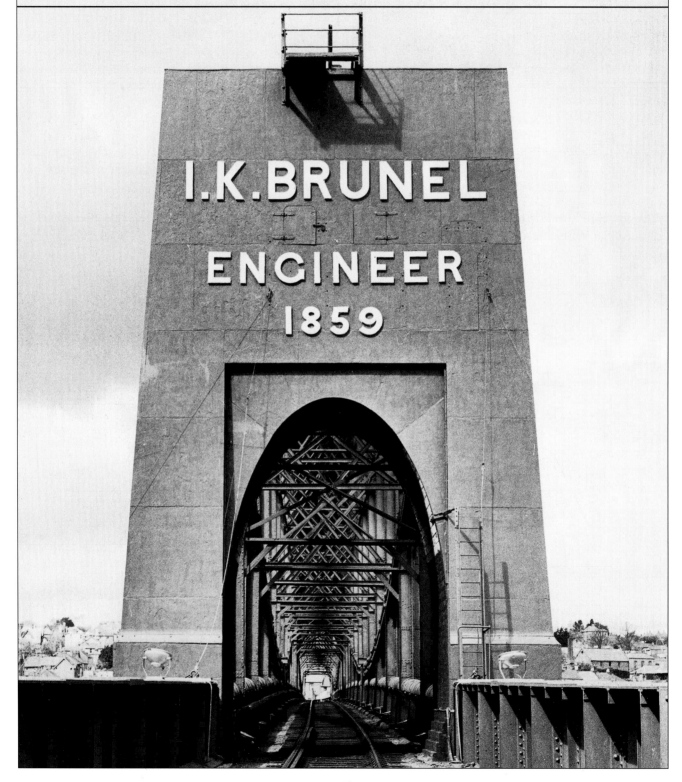

The extension of Brunel's broad-gauge network into the far west of England was a slow process, not helped by the fact that two rival schemes to link Cornwall with the rest of the country were proposed initially. One proposal was what was known as the 'Central line', which ran inland on the old coach route, and the other a more southerly line, which would link up with the South Devon Railway at Plymouth. The debate over the choice of route lasted for over a decade, and need not be discussed here, other than to say that considerable argument still raged in Cornwall in 1843, when Capt William Moorsom was asked by local promoters to survey a route for a railway through Cornwall. By then the promoters of this railway had gained the support of the Great Western, so it was probably no surprise that, when the Cornwall Railway Bill was presented to Parliament in 1845, the supporters of the central route continued to press their case by submitting a Bill for the Devon & Cornwall Central Railway, supported by the London & South Western Railway.

The bill which the LSWR supported was thrown out of Parliament on a number of grounds, not least the huge engineering difficulties which would have been faced by its promoters. It was noted that scarcely any stretch of the line would have been built on natural ground, viaducts, bridges, tunnels and cuttings being needed to drive the route through the hilly landscape. Moorsom's route was not completely satisfactory either, containing as it did ferocious gradients of 1 in 60 over long stretches of line, and many sharp curves,

largely because it was intended that the railway should use the atmospheric system. To cross the River Tamar, Moorsom proposed a train ferry at Torpoint. Although passed by the House of Commons, the Cornwall Railway Bill was rejected in the House of Lords, its members indicating that they might take a more sympathetic view if a better scheme were submitted which reduced the gradients and removed the need for a ferry. Brunel was then asked by the Cornwall Railway promoters to undertake a new survey of the line. This he did, proposing a railway which crossed the Tamar by means of a high-level bridge at Saltash, and followed much the same route as that originally designed by Moorsom, although with much gentler gradients. Moorsom was, not surprisingly, unhappy at being usurped by Brunel, but had little choice but to accept the inevitable. The Cornwall Railway Bill went back to Parliament, and was eventually passed in August 1846. Some work began on the new line, but stopped at the end of 1847 due to lack of funds. In 1851 Brunel suggested that, by reducing the railway to single-track only, costs could be sufficiently reduced to allow the whole scheme to be completed for around £800,000, a sum which would include the bridge over the Tamar at Saltash. Capital was still short, however, and it was not until the following year that the company was finally in a position to recommence construction of the line.

Whilst work had been suspended on the Cornwall Railway, Brunel had been fully occupied on a number of other projects, including the construction of his new station

Left: Brunel's name was added to the towers of the Royal Albert Bridge as a tribute to his genius after his death in 1859. *British Railways*

Above: One of Gooch's Great Western locomotives, *Lord of the Isles*, took pride of place within the displays at Hyde Park. This view was taken from the *Illustrated Exhibitor. Illustrated Exhibitor*

at Paddington, which was mentioned in Chapter 3. He had also been heavily involved in the planning of the Great Exhibition, which had opened in Hyde Park in May 1851. Although the initial idea for the exhibition came from Prince Albert, much of the success of the venture was down to the enthusiasm and drive of Henry Cole, a civil servant who had seen the economic benefits brought to France by the Paris Exposition of 1840. There were many objections to the whole idea, but a Royal Commission chaired by Prince Albert brushed aside criticism and doubts, and within two years the exhibition had been planned, built and opened. Brunel was asked to be a member of the Building Committee, and was chairman of the judges for the 'Civil Engineering, Architecture and Building Contrivances' section of the exhibition. He was also a member of the Machinery Section Committee. Perhaps his most important role was as a member of the Building Committee, which was given the difficult task of deciding on the design of the building to hold the great event. There were 245 entries submitted, none of which was deemed suitable, and the Committee set about producing its own plans. The end result was even less satisfactory — a low brick-built structure which drew a good deal of public criticism. The building was topped with an immense iron dome 150ft high, designed by Brunel, who is reported to have objected to the whole idea of a permanent building for the purpose of such an exhibition anyway. The building should be in the 'Railway Shed Style', he argued, and, when Joseph Paxton submitted his own plans, Brunel was one of his strongest supporters, arguing that the design put forward was 'the best adapted in every respect for the purpose for which it was intended'. The iron and glass building produced by Paxton, which became known as the Crystal Palace, was planned and built within a year, with much assistance from Brunel. It is not surprising, therefore, that Brunel's new station at Paddington should be based so closely on this magnificent structure. When the Crystal Palace was moved to Sydenham, Brunel was once again heavily involved in the project, designing two large water-towers for the site.

Back on the Cornwall Railway, work had first restarted on the 15-mile stretch between Truro and St Austell, but by the latter part of 1852 contracts had been placed for the section between Plymouth and Saltash. Brunel's original plan for the bridge over the River Tamar at Saltash had been for a timber structure, consisting of one main span of 255ft, with six further spans, each of 105ft. However, this scheme was rejected by the Admiralty, which insisted that at high tide there should be headroom of at least 100ft for vessels. The great engineer had also considered the idea of a bridge with one span of over 1,000ft, which would avoid the need for a central supporting pier, but this idea came to nothing, chiefly due to the enormous cost, which he estimated at £250,000

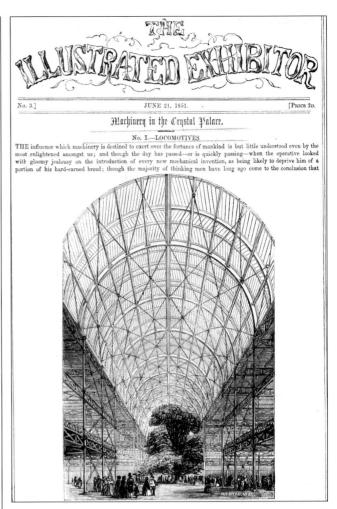

THE ILLUSTRATED EXHIBITOR

No. 3.] JUNE 21, 1851. [PRICE 2D.

Machinery in the Crystal Palace.

No. I.—LOCOMOTIVES

THE influence which machinery is destined to exert over the fortunes of mankind is but little understood even by the most enlightened amongst us; and though the day has passed—or is quickly passing—when the operative looked with gloomy jealousy on the introduction of every new mechanical invention, as being likely to deprive him of a portion of his hard-earned bread; though the majority of thinking men have long ago come to the conclusion that

Above: A view of the interior of the Great Exhibition, as seen in the *Illustrated Exhibitor* for June 1851. The similarities with Brunel's later Paddington station are striking. *Illustrated Exhibitor*

for the bridge and another £250,000 for supporting works. Clearly the already shaky finances of the Cornwall Railway could not stand such expense, and thus Brunel's final design was for a bridge with two main spans, each of 455ft, supported by a central pier.

Before work could begin, Brunel made a very careful survey of the river bed to ascertain the best place to build the foundations for the central pier. At the outset it seemed that there was no suitable location for this, since there was 70ft of water over a layer of thick mud. In order that the bed of the river could be properly surveyed, in 1849 Brunel designed a large, wrought-iron tube, which was 85ft long and 6ft in diameter. Using lifting-tackle attached to two hulks this was sunk through the mud and clay on the river-bed to the solid rock below. No fewer than 175 holes were made, allowing an accurate survey of the bedrock to be completed. A model of the river bed was then made, and the site for the central pier chosen. Lack of finance prevented further work being done,

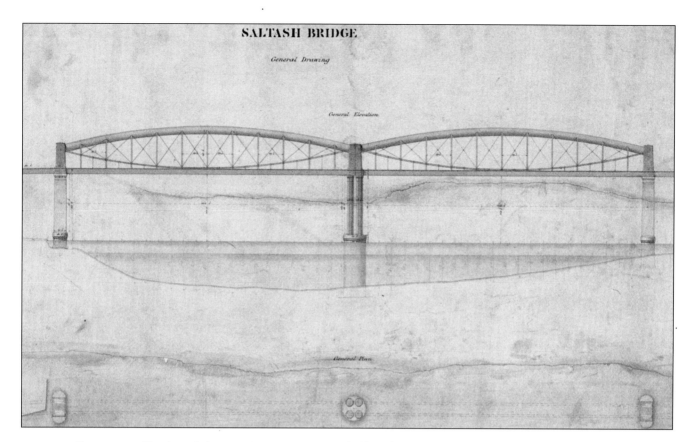

SALTASH BRIDGE

General Drawing

General Elevation

General Plan

Above: One of Brunel's original drawings showing the general arrangement of the Royal Albert Bridge at Saltash. *Railtrack Great Western*

save a small amount of brickwork being constructed on the river bed to check if Brunel's designs for the pier would work.

Work did not recommence until 1853, and during the hiatus every effort was made to reduce costs. By far the biggest saving was made when it was decided to reduce the bridge to a single-track structure. Further savings were made for the Cornwall company when Brunel was able to negotiate with contractor C. J. Mare a very low price, £162,000, for construction of the bridge. To build the central supporting pier, Brunel devised a larger tube which he called the 'great cylinder'. This 90ft-long tube was constructed in two sections; at the bottom was a diving-bell, to which compressed air could be supplied. A 10ft-wide pipe ran up the centre of the tube, and this contained both an airlock and pumps to remove any water that might appear during excavations. Weighing over 300 tons, this apparatus was constructed on the riverbank, and was floated out to midstream in May 1854. Attached to four pontoons, the tube was manoeuvred into position, and water slowly pumped inside, allowing it to float upright. Once this had been done, the entire structure was allowed to sink to the river bed, the whole process taking around two weeks.

The detailed survey carried out in 1849 had shown that the river bed sloped, and so one side of the base of the tube was 6ft longer than the other; when it sank into the mud, problems were encountered as it struck a large bed of oyster-shells, and some irregularities on the river bed. As a result the cylinder rested at an angle, and, after heavy weights had been loaded on to one side of the tube, men climbed into the pressurised bell at the bottom to excavate the obstructing rock. Once this had been done, the tube sank into position and finally stood vertically on the rock. The rock itself was a very hard 'greenstone trap' which proved extremely difficult for the workmen to dress and excavate; eventually a ring of granite masonry was constructed, despite the influx of water which had constantly to be pumped out.

Work slowly continued, and by the end of 1856 the masonry pier was complete and it was possible for the great tube to be dismantled and floated back to the shore, its work done. Whilst the workmen had bravely built the masonry for the pier, deep under the Tamar, work had continued apace on the supporting columns for the 17 side-spans, as well as on the two other main piers supporting the bridge at each end. The two main bridge-trusses were constructed on the Devon side of the river, and each consisted of an arched wrought-iron tube, which was oval in cross-section. Attached to each tube were suspension chains, connected to each other by 11 uprights, to which the longitudinal girders carrying the railway track were fixed. Further diagonal bracing was added to give some rigidity to the whole structure. Described as a combination of an arch and a suspension bridge, this arrangement was a further development of the ideas Brunel had pioneered at Chepstow, although (as already mentioned) he had constructed a very much smaller bridge using wrought-iron tubes to cross the docks at Bristol, some years before the building of the bridges at Chepstow and Saltash.

The two trusses weighed 1,060 tons each, and the task of placing them in position was not an easy one. As well as gaining experience at Chepstow, Brunel had also assisted Robert Stephenson at the lifting of the Conway and Britannia bridges; also present at all three of these operations was Brunel's old friend, Christopher Claxton, who had supported him in the construction of the *Great Western* and *Great Britain*. Meticulous planning was undertaken, and the arrangements tested and discussed before operations commenced, a luxury Brunel was not to enjoy some months later when the *Great Eastern* launch was undertaken. Having excavated a dock in the riverbank near the end of the bridge-truss, two large, iron pontoons were floated into position and then sunk on to timbers positioned underneath; a timber, supporting framework was then built on top of the pontoon. On 1 September 1857 the much-rehearsed operation took place, beginning with the western truss, and watched by an estimated 300,000 people gathered on the sides of the river. To ensure that everything worked well, around 500 workmen were stationed at various locations, with Captain Claxton not surprisingly in charge of 'arrangements afloat'. Brunel was also assisted by his engineer, Robert Brereton, and Captain Harrison, the Master of the *Great Eastern*. The huge truss having been winched into position on the pontoon, water was pumped out and, once it had been raised by the tide, Brunel gave the signal for the whole assembly to be gently floated out into the river. Positioned on a temporary platform situated on top of the truss, Brunel directed operations, and the movements of the pontoons and winches were controlled by a series of flag signals until the truss was in position over the bases of the western and central bridge piers. Water was then pumped into the pontoons, which were floated away, allowing the truss to rest on the pier bases.

Rather than construct the whole of the supporting bridge pier in advance, Brunel had decided instead to build the pier under the truss. Three hydraulic jacks were positioned under the end of each bridge-truss, and raised 3ft at a time. Masonry was then constructed under the jacks, which had an additional safety feature in that each ram had a screw-thread, on to which was fixed a locking-nut which was kept tightly screwed to the top of the jack at all times. Great care was taken to allow the cement to set, but by May 1858 the first

Right: A view of the bridge during construction, with the trusses about to be lifted into position. *Swindon Museum Service*

truss had been raised 100ft to its correct position. Within two months the eastern truss was floated into position in a similar operation to that carried out the previous year, this time supervised by Brereton, since Brunel was in Europe convalescing; by December this second truss was also in position. Once again Brunel was absent due to his deteriorating health, and it was left to Brereton to supervise much of the work needed to complete the bridge.

Regrettably, Brunel was also unable to witness the opening of the bridge on 2 May 1859, when it was formally commissioned and named 'The Royal Albert Bridge' by the Prince Consort. A few days after this event, Brunel visited Saltash and was drawn across the bridge, lying on a couch placed on a railway wagon — a rather tragic image of a man broken by work on his other great obsession, the *Great Eastern* steamship, which had occupied much of his time and thoughts during the design and construction of the bridge. Brunel's son noted that, although striking in design, the bridge had little in the way of adornment, its simple lines and design an embodiment of engineering skill. After the engineer's death, the inscription 'I. K. BRUNEL ENGINEER 1859' was added by the Cornwall Railway to the landward archway of each bridge pier as a memorial to

his work, and despite some strengthening work over the years, the bridge still stands today as a testament to his design flair.

Although the Royal Albert Bridge was perhaps the most dramatic engineering solution Brunel was required to produce on the Cornwall Railway, the landscape the railway was to cross required much thought, since it included eight river estuaries and well over 30 valleys which would have to be bridged. To achieve this within the constraints of the Cornwall Railway's chronic cashflow problem, Brunel elected to construct a series of timber bridges, most of which would be supported by masonry piers, although, where bridges crossed tidal creeks, timber piles were hammered into the mud to support the structure. Brunel's first timber bridge had been a very small affair, consisting of five spans which crossed Sonning Cutting, near Reading, and carried a small public road. It is likely that this structure was used by Brunel as a prototype for his later designs, and further examples were constructed on the Cheltenham & Great Western Union Railway, at Stonehouse, Bourne and St Mary's. Although

Below: The construction of the bridge is well advanced in this stereo-view by William May taken in 1859. *Royal Institution of Cornwall*

BRISTOL & EXETER RAILWAY.

VISIT

OF HIS ROYAL HIGHNESS

THE PRINCE CONSORT,

TO THE

OPENING

OF THE

ROYAL ALBERT BRIDGE,

AT

SALTASH,

ON

MONDAY, 2nd May, 1859.

ROYAL TRAIN TIME BILL.

DOWN	DEP. A.M.	ARR. A.M.	UP	DEP. P.M.	ARR. P.M.
WINDSOR	6 0		SALTASH	—	
Bristol		8 35	Cornwall Junction		—
"	8 45		"	6 50	
Taunton...		9 35	Newton		—
"	9 38		"	—	
Exeter		10 25	Exeter		8 15
"	10 35		"	8 25	
Newton		11 5	Taunton...		9 12
"	11 10		"	9 15	
Cornwall Junction		12 0	Bristol		10 5
"	12 5		"	10 15	
SALTASH		12 15	WINDSOR		12 50

The following arrangements will be necessary for the proper working of this Train, which must be strictly attended to:—

The 7.50 a.m. Down Passenger Train is to Shunt at Tiverton Junction.

The 8.0 a.m. Goods Train Down will not start from Bristol until after the Royal Train.

The 8.0 p.m. Up Train is to Shunt at Tiverton Junction.

The 9.20 p.m. Short Train from Weston is to Shunt at Yatton.

BRISTOL, 29th April, 1859.

Left: Poster of the opening of the bridge. *Swindon Museum Service*

Above left: A view of the bridge in broad-gauge days, with Saltash station in the foreground. *Swindon Museum Service*

Left: Brunel's Saltash Bridge seen during the conversion of the broad-gauge in May 1892. *Illustrated London News*

Above: A driver's eye view of the Royal Albert Bridge gives a cathedral-like appearance to the mighty trusses. *British Railways*

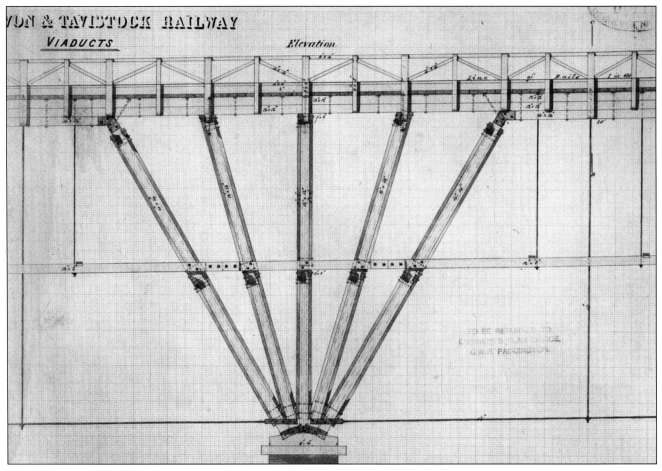

these bridges were relatively small, four rather more impressive structures were built for the South Devon Railway, including the Ivybridge Viaduct, which consisted of 11 spans of 61ft, supported by masonry piers which featured twin tapered columns rising to rail level. An even larger bridge was built at Landore, on the South Wales Railway, which was 1,760ft in length, and had 37 spans of differing dimensions, the supporting trusses being a mixture of timber and masonry.

In the case of the Cornwall Railway, Brunel utilised timber viaducts on a scale seen nowhere else on his railway system; over the 60 miles of railway, 42 bridges were constructed, their total length being over four miles. Although cheap to build, it was later noted that the additional maintenance required probably meant that it would have been more economical to build masonry bridges in the first place! In general the viaducts were built using a standardised system, with uniform dimensions for spans and component timbers. Three beams ran longitudinally, supporting the roadway on which rails were laid, and these beams were themselves supported by four timbers radiating out from the tops of the masonry piers. The whole structure was further strengthened

Left: The Great Western photographer has framed a wonderful photograph showing Brunel's Treviddo Viaduct in Cornwall. The picture is dated 1895. *National Railway Museum*

Below left: An original drawing of a timber viaduct for the South Devon & Tavistock Railway. *Railtrack Great Western*

Below: Another 'official' Great Western photograph, this time of Tregagle Viaduct, which shows very clearly the robust and basic way in which these viaducts were constructed. *National Railway Museum*

with diagonal bracing and wrought-iron tie-bars. The timber, which was largely Baltic pine, was treated by means of the 'Kyanizing' process, used extensively by Brunel on his 'baulk timbers' for broad-gauge track, whereby the wood was dipped in tanks containing a solution which reduced rot and decay. Nevertheless, much care was needed with these slender and elegant bridges, and constant inspection took place in order that any faults could be quickly detected. Including other, similar bridges built by Brunel for the West Cornwall Railway, there were a total of 49 timber viaducts constructed in Devon and Cornwall; as well as the high cost of maintenance, steadily increasing loadings, due to heavier locomotives and rolling stock, meant that bridges were gradually replaced. Although this process began in the 1870s, the last in Cornwall, at College Wood on the Falmouth branch line, was not replaced until 1934, and another example in South Wales lingered until after World War 2.

It is perhaps typical of Brunel's character that, less than four years after the Dundrum Bay fiasco, and around the time that the Great Western Steamship Co was formally wound up, his thoughts should once again return to the field of ship design. The impetus came from a request by William Hawes, Chairman of the Australian Royal Mail Company, for Brunel to act as consultant engineer to the new shipping line. This had been set up to run a regular service to the colony, which was becoming increasingly important due to the discovery of gold in the late 1840s. The company required two fast steamships for the route, and Brunel's general specification called for a vessel of between 5,000 and 6,000 tons, which would only need to refuel once, at Cape Town. For the detailed design and construction of these ships, Brunel turned to John Scott Russell, who was asked to submit a bid to the steamship company. Russell was a distinguished naval architect, who, whilst Secretary of the Royal Society of Arts, had been a key figure in the promotion and opening of the Great Exhibition of 1851. He was also a co-owner of the Crystal Palace Co, set up to rebuild that great structure at Sydenham. No doubt it had been through Scott Russell that Brunel had been given the job of designing two water-towers for the Crystal Palace at its new site, although the two engineers had known each other for some considerable time, possibly as early as 1836.

For the most part, the two ships built to Brunel's specification were of Scott Russell's own design. The ships, later named the *Adelaide* and the *Victoria*, each featured his characteristic 'wave-line' hull and had watertight bulkheads throughout. Another innovation was the use of what we

Left: When the timber in viaducts became life-expired, the Great Western replaced them with either new bridges or embankments if practical. This view shows the replacement of the viaduct at St Austell. *National Railway Museum*

Above: Although in many cases Brunel was able to design an elegant solution to an engineering problem, this was not always the case. The timber bridge over the River Thames at Shiplake, on the Henley branch, opened in 1857, is a far cry from the graceful arches of the Maidenhead Bridge on the same river. *National Railway Museum*

would now call 'box-girders' running the length of each ship, which gave them great longitudinal strength. Russell owned a shipyard at Millwall, and it was here that the two ships were constructed; Vaughan records that both were launched within six months of their keels being laid. By 12 December the *Adelaide* had been completed and set sail for Australia, completing the trip in two months. Perhaps the successful refitting of his beloved *Great Britain* for the Australian route may also have contributed to Brunel's renewed interest in ship design; his son noted that, in late 1851 and early 1852, he spent much time on the idea of 'a great ship' capable of running a service to India or Australia. Writing a few years later, Brunel himself noted that 'I matured my ideas of the large ship with nearly all my present details, and in March [1852] I made my first sketch with paddles and screw'.

The 'great ship' proposed by Brunel was certainly large; at 600ft in length, she would be far bigger than anything then afloat, and almost double the length of the *Great Britain*. Brunel collaborated further with Scott Russell, as well as discussing the idea with Christopher Claxton and a few other friends, and gradually the basic design of what became the *Great Eastern* began to evolve. Scott Russell's experience as a shipbuilder led him to confirm that a ship built to Brunel's designs would have a displacement of between 18,000 and 20,000 tons, and was likely to require an engine of 850hp to maintain an average speed of 15 knots. After further discussion, Brunel also decided that a combination of screw propeller and paddle-wheels should be used. There was little doubt that a ship of such a size would be pushing at the boundaries of existing technology in many areas, something which no doubt attracted Brunel greatly, but, having agreed the basic concepts of this monster, Brunel and Scott Russell had to find backers to fund it. They did not need to look too far, for at Scott Russell's suggestion, an approach was made to the directors of the Eastern Steam Navigation Co, a shipping line set up in January 1851 to compete for mail contracts between England, the Indian subcontinent, China and Australia. The hopes of the fledgling company were dealt a severe blow in March 1852, when the government awarded the contract to the Peninsular & Oriental line, despite the fact that the new company had tendered a bid lower than P & O's.

Left without a real purpose, the Eastern Steam Navigation Co was thus open to suggestions for new ventures. In June 1852 Brunel wrote a short report to the directors, outlining his ideas for the new ship; he noted with no little understatement that there was little that was very novel about the scheme, and that what was proposed was only 'to build a vessel of the size required to carry her own coals for the voyage'. The size of the ship, he concluded, was limited only by the extent of demand for freight, and by the circumstances of the ports at which she would call. The company appointed a committee to examine the proposals for this new ship, and it was arranged that Brunel should meet its members in July 1852. On the day set for the interview with the directors, Brunel was ill, and John Scott Russell appeared in his place. This may well have been a good decision in the circumstances, for, as well as being an eloquent public speaker, his reputation as a shipbuilder led many board members to take note of his arguments. There is little doubt that Brunel's reputation as something of a maverick would have counted against him, had he appeared, despite his own persuasive abilities as an orator.

Even with the reassurances given by Scott Russell, some committee members, including the chairman, were not convinced and resigned from the board. In view of what was to become of the company in the future, this may well have been a wise move, and one can only marvel at the confidence of those who put their faith in what must have seemed a risky proposition. Some, no doubt, were seduced by the prospect

Below: The screw engines of the *Great Eastern*. Like the paddle engines, these were the largest then built. They were of such quality that the *Illustrated London News* claimed that: 'The *Great Eastern* would be the most complete steam vessel afloat.' *Illustrated London News*

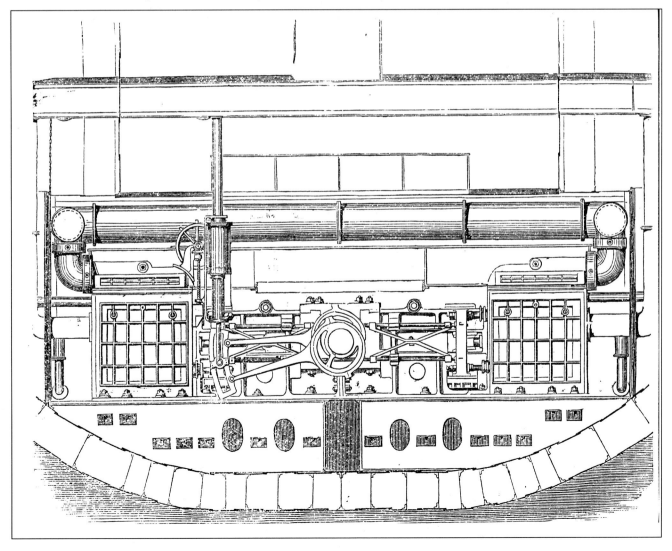

of the profits such a ship could earn; Brunel was quick to point out that, if goods could be carried direct in 30-35 days, 'the certainty of freight ensures a return far beyond all present proportion of return to cost'. Brunel was well enough to meet the directors a few days after the interview with Scott Russell, and was unanimously appointed as Engineer for the project. The company determined that work could start when £800,000 capital or 40,000 shares had been invested. Since some of the directors had resigned, more investors were needed, and Brunel persuaded a number of influential figures, such as Samuel Peto and Thomas Brassey, the railway contractors, to purchase shares, as well as Charles Geach, Chairman of the Midland Bank, and business partner and friend of John Scott Russell. Brunel himself had a substantial stake in the company, buying 2,000 shares, although it is obvious that his investment went far further than mere financial considerations. Writing to the directors some months later, he noted that he felt an enormous responsibility, bearing in mind that he had 'induced more

than half the present Directors of the Company to join', and that he had 'never embarked in any one thing to which I have so entirely devoted myself'.

Before construction could start, more work was necessary. Writing to Scott Russell on 13 July 1852, with regard to the sheer size and scale of the vessel proposed, Brunel confidently stated that 'the wisest and safest plan in striking out a new path is to go straight in the direction which we believe to be right, disregarding small impediments which may appear in our way'. Presumably there was much discussion between the two engineers regarding the form and features of the new ship, since Brunel then reported to the Eastern Steam Navigation board on 21 July on the 'Mode of Proceeding'; in this he noted that he had been engaged 'very constantly' in developing detailed plans for the new ship, and that the decision to proceed further with the project was a great one and must be taken with 'deliberation and certainty'. With regard to the construction of the ship, he continued, 'nobody can, in my opinion, bring more scientific and practical knowledge to bear than Mr Scott Russell'. Also mentioned were the two companies with which Brunel had a great deal of discussion with regard to the engines to be used in the great ship, Maudslay, Son & Field and James Watt & Co.

Below: The engine room for the paddle engines. Each of the cylinders measured 6ft 2in with a 14ft stroke. *Illustrated London News*

There was a considerable delay whilst further discussion and research was carried out, before tenders could be invited and considered. It was thus not until May 1853 that Brunel was able to report to the board on the submissions received from the various tenderers, and make recommendations. It was not perhaps surprising, bearing in mind the relationship built up with Brunel, that the tender submitted by Scott Russell was the one recommended, with Scott Russell proposing to build and fit out the ship, as well as provide the paddle-engines and boilers for £275,200, with a lower figure of £258,000 being proposed if a second ship were built to a similar design, as Brunel had originally intended. The only remaining contract was for the engines for the screw propeller, for which a tender from James Watt & Co was accepted. In the circumstances, the cost quoted by Scott Russell was hopelessly over-optimistic; bearing in mind the enormous size of the new ship, and some of the new features to be incorporated, it is difficult to understand how such an experienced shipbuilder could have thought that the job could be done so cheaply. Brunel himself had originally estimated the cost of the new ship to be nearer £500,000, but since the project was struggling to raise enough finance anyway, he was no doubt happy to accept a lower figure.

Once the tenders had been accepted, there was a further delay of almost six months whilst contracts were negotiated. The steamship company still found it difficult to raise the capital required to allow work to begin on the new ship, and Charles Geach, Scott Russell's partner, was called upon to assist when a number of shareholders, perhaps nervous at the prospects for the new venture, refused to continue. Matters were not helped either by a serious fire at Scott Russell's yard, which caused over £150,000 of damage. Once again, Geach came to the rescue, allowing his partner to stay in business. On 22 December 1853 contracts were finally signed, giving Scott Russell responsibility for the detailed design and construction of the ship, but allowing Brunel 'entire control and supervision over the proceedings and workmanship'. For someone not used to working closely in partnership with other engineers, this arrangement seems, with hindsight, to have been somewhat naïve, and the confidence with which Brunel recommended Scott Russell to the board was to be short-lived.

A pressing problem for the engineers was the method by which the new ship could be launched, since, when constructed, she would be far too big to fit within Scott Russell's own yard. Scott Russell had suggested the construction of a brand-new dry dock in which to build the vessel, which would have allowed her to be floated out when completed, but Brunel, having considered the proposal, rejected it both on the grounds that Scott Russell's original estimate of around £15,000 was hopelessly optimistic, and that the treacherous sand and gravel of the River Thames, of which he had gained much experience whilst working on the Thames Tunnel, would not be suitable when excavated. Land was available in the yard of David Napier & Co, next to Scott Russell's shipyard on the Isle of Dogs, and so this was leased to allow other work to continue.

The size of the ship presented a further difficulty, in that launching her end-on into the Thames would require the bow to be raised to a considerable height. Notwithstanding the practicalities of supporting a hull weighing over 8,000 tons, and constructing it on a considerable gradient, there was also the problem that, once fitted with engines and other heavy equipment, the hull might be damaged or strained when actually launched. Brunel thus decided to launch the new ship side-on, and a site was chosen above the high water mark, where a vast, inclined platform consisting of oak piles was driven into the riverbank. Over 12,000 tons of wood was used, the bottom of the ship resting on piles which protruded 4ft above the ground. At this point Brunel had not yet worked out in any detail how the ship would be launched from this position into the river, but hoped that some kind of 'patent slip' using rollers could be built.

Work finally started on the vessel in the spring of 1854, and slowly she began to take shape. The ship was to be 680ft long, with a beam of 83ft; to give a vessel of this length the kind of longitudinal strength required to cope with heavy seas, Brunel and Scott Russell adopted a new hull design; this featured an inner and outer skin, between which were sandwiched girders running the length of the ship. As with the *Great Britain*, a cellular form of construction was adopted, with the ship divided internally by a number of watertight bulkheads. Two further longitudinal bulkheads in the centre of the ship separated the engine and boiler compartments. Construction began with the keel, bulkheads and inner skin. The plates which made up the hull were rolled to three standard sizes, with thicknesses ranging from ½in to 1in, and were cut and shaped according to wooden patterns. A model of the ship was used, marked with the precise position of each numbered plate; once the holes for rivets were punched out, each plate was transported by wagon to the area where it was to be used. Brunel had rejected Scott Russell's offer to construct a travelling crane capable of lifting up to 60 tons, considering it 'very frightening', and thus the workmen had to haul these heavy, 10ft by 2ft 9in iron plates into position, using only block and tackle equipment. The plates were then riveted in place by a squad of workmen which included two riveters, a 'holder on' and two boys whose job it was to heat and place the rivet in the hole. Over 3 million rivets were used in the construction of the hull, to hold over 30,000 plates.

The engines were also much larger than anything that had gone before. If there had been amazement at the size of the crank-axle for the *Great Britain*, those required for Brunel's new ship were to cause even greater consternation. The paddle engines were the largest ever built, and the crankshaft for the new vessel was eventually forged at the Glasgow firm of Fulton & Neilson, a process requiring new, much larger steam hammers. All manner of new equipment was needed, since nothing on this scale had been attempted before; as well as suppliers and manufacturers away from the shipyard, Scott Russell himself had to invest in much new plant to allow work to proceed, including a lathe large enough to machine the crankshaft.

Towards the end of 1854, two events occurred which presaged future problems for the great ship. The first was a deterioration of relations between Brunel and Scott Russell, a situation which had a good deal to do with the contractual arrangements already mentioned, whereby Brunel maintained overall control in the project, despite the fact that Scott Russell was ultimately responsible for completing the work. Brunel's attention to detail and his insistence on making alterations to plans and designs as work continued, coupled with the fact that much of what was being done was new, led to a slowing of progress on the ship, which, Brunel conceded, was unlikely to be launched within the 18 months originally specified. Relations between the two men were not helped by a newspaper article in the *Observer* of 13 November 1854 which, although containing many inaccuracies regarding the construction of the ship (now named *Leviathan*, because of her enormous size), and other matters regarding iron steamship construction, made only brief reference to Brunel's involvement, noting that 'Mr Brunel...approved of the project, and Mr Scott Russell undertook to carry out the design'.

There seems little doubt that Brunel suspected Scott Russell's complicity in the anonymous article and, in a letter to the Secretary of the Eastern Steam Navigation Co a few days later, claimed that the article 'bears rather evidently the stamp of authority, or at least it professes to give a great amount of detail which could only have been obtained from ourselves'. Brunel was 'much annoyed' by the article, claiming that it deprecated 'the efforts which I had successfully made in advancing steam navigation'. He concluded his letter by stating that, whilst having responsibility for the success of the project, he could not countenance the idea put forward in the article that he was a 'mere passive approver of the project of another, which in fact originated with me'. The vehemence of Brunel's letter reveals more than a little professional jealousy, and it must be noted that, although Brunel had originated the idea for the new ship, he would not have been able to advance his plans without the valuable assistance of Scott Russell and others.

Brunel's growing frustration was not helped by the increasingly fragile state of Scott Russell's finances. In the same month that the infamous *Observer* article was published, Charles Geach, Scott Russell's partner and major backer, died suddenly, leaving him with a serious cashflow problem. Clearly over-extended, he had much of his capital tied up in shares in the steamship company — shares which had little value. In early 1855 new financial arrangements were made with Scott Russell's bankers to pay him in instalments, subject to a certificate being signed by Brunel when work had been completed by Scott Russell. However, the situation worsened as the year progressed, and in April the method of launching the ship became another source of dissension. Brunel had been forced to abandon the idea of using some kind of roller arrangement, as the directors had deemed this too costly; whilst he had not decided on the exact form the sideways launch would take, he was determined that it should be controlled, allowing the ship to be moved down to the low water mark where it could be floated on the tide. Scott Russell, however, was in favour of a free launch, largely because he was contracted to launch the vessel, and Brunel's preferred option was clearly more expensive, although he was also aware that the friction generated by a ship of 12,000 tons on a greased slipway would itself act as a controlling measure in any free launch.

In the midst of the construction of Saltash bridge and his new steamship, Brunel involved himself in two markedly different projects of national importance. The feeling of national pride and enthusiasm occasioned by the Great Exhibition of 1851 had swiftly evaporated during the course of the Crimean War, which had started in September 1854 and was destined to continue for 18 long months. Whilst the courage of the British Army was widely acknowledged, the ineptitude of the government, and of the War Office in particular, had caused a public outcry. In 1855 Brunel had submitted a design for a floating 'gun-carriage' which could be used to shell some of the Baltic ports; the guns were to be propelled by jets of steam, but nothing came of the idea, bureaucrats in the War Office doing little until it was too late, and the war had ended.

Far more significant was Brunel's involvement in the design and construction of a prefabricated hospital for the Crimea. Conditions in the main hospital for British troops at Scutari were appalling, with poor sanitation and an untreated water supply making it a national disgrace. Largely through the efforts of Florence Nightingale, the awful conditions at Scutari became public knowledge in England. The government's handling of the war led to its downfall in January 1855, and the new administration headed by Prime Minister Lord Palmerston set up a Sanitary Commission to investigate conditions in the Crimea. A month later, Brunel received a letter from the Permanent Undersecretary at the War Office, Benjamin Hawes, asking him to design a hospital which could be built in England and shipped to the Crimea very quickly. On the face of it, this request seems

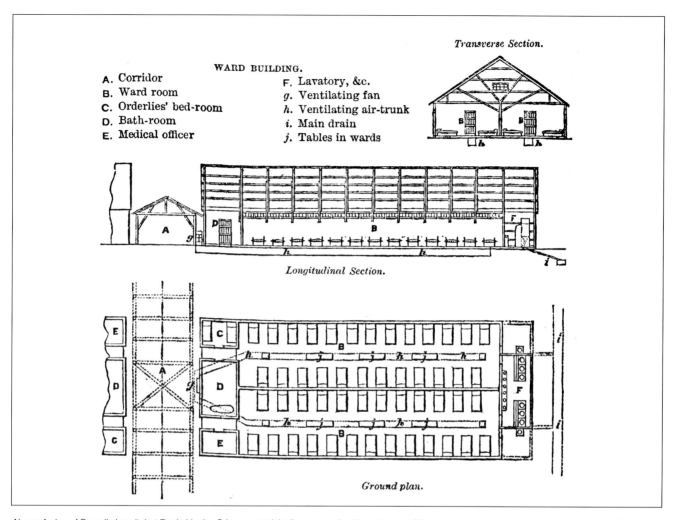

WARD BUILDING.

A. Corridor
B. Ward room
C. Orderlies' bed-room
D. Bath-room
E. Medical officer

F. Lavatory, &c.
g. Ventilating fan
h. Ventilating air-trunk
i. Main drain
j. Tables in wards

Transverse Section.

Longitudinal Section.

Ground plan.

Above: A plan of Brunel's hospital at Renkoi in the Crimea, as originally reproduced in the biography written by his son in 1870. **A**uthor's Collection

perfectly straightforward; however, it should be noted that Benjamin Hawes was Brunel's brother-in-law, and a figure despised by Florence Nightingale for his arrogance and inefficiency. Hawes had managed to maintain his position despite all the changes made by the new government, and there is little doubt that he and Brunel had already discussed the hospital scheme some time before the official letter was sent. This seems all the more likely since, within six days, Brunel had been able to submit plans to the War Office for a 1,000-bed hospital, which consisted of a number of standardised buildings, each with two wards of 24 beds. Much attention was paid to the ventilation, washing and sanitary arrangements of the hospital, and, once plans had been approved, Brunel meticulously planned the movement of the raw materials, tools and labour needed to erect the buildings. John Brunton, an assistant working for Brunel, was chosen to travel to the Crimea to supervise the building, and by 12 July 1855 the first 300 beds were ready for use. The site chosen for the hospital was not Scutari, but Renkoi.

The War Office continued to demonstrate its inefficiency, since no troops were moved to the hospital until 2 October; by Christmas over 1,000 beds were available, and although the war ended a month later, it was nevertheless reported that, by May 1856, 1,331 patients had been treated, of whom only 50 had died. Important though Brunel's contribution had been, his relationship with Hawes appears full of contradictions; although they were close friends, Hawes stood for much that Brunel himself despised, particularly government inefficiency and interference. Clearly the bond of friendship was strong, for Brunel seems to have supported Hawes, and in so doing excluded Florence Nightingale, who, through her exposition of the terrible conditions at Scutari, had made few friends in high places. At no point is she mentioned in any of the correspondence between Brunel and the War Office, despite the fact that many of her principles of hygiene and cleanliness were incorporated into Brunel's designs for the hospital. Apart from the personal animosity between her and Hawes, her isolation in the planning and execution of Brunel's designs may well also have reflected the prevailing attitude towards the role of women.

Meanwhile, in Millwall, the deterioration of the once cordial relationship between Brunel and Scott Russell was demonstrated by a change in the tenor and content of correspondence between them. Rolt notes a cooling marked by a more formal style of letter; as the year progressed, Brunel's correspondence became more hectoring and irritated, perhaps exacerbated by his increasingly poor health. Admonishing Scott Russell for the lack of accurate information on the weight of the ship, he wrote: 'I wish you were my obedient servant, I should begin with a little flogging.' Following Scott Russell's repeated requests for further payment, matters finally came to a head in February 1856, when Brunel advised the steamship company to take possession of the ship since, he argued, Scott Russell was in breach of his contract. Whilst not actually bankrupt, he was in no position to continue, and Brunel was then forced to take charge of the operation himself. This was not, however, the end of Scott Russell; in a bizarre twist to the story, he was asked by the liquidators to produce estimates for the completion of the ship. This he did, although work on the ship only restarted in May 1856, with Brunel in sole charge, reluctantly approving an arrangement whereby Scott Russell was nominally one of his assistants. It is perhaps significant that, for a period in June, when Isambard was ill and unable to interfere in the more minor affairs of the shipyard, fairly rapid progress on the ship was made by Scott Russell and his assistants. Brunel was once again unhappy that his authority had been undermined, and that decisions had been made without his approval. Some months later Brunel complained formally to the board, with Scott Russell arguing that he had kept the engineer fully informed of progress, and had dealt with all his instructions courteously. Having informed the directors that he would be taking a holiday anyway, Scott Russell took his leave of the project and did not return, leaving Brunel in full control.

Managing the completion of the ship was not easy, and it seems clear that Brunel soon realised the full enormity of the task, a situation hardly helped by his failing health. With his insistence on controlling every aspect of the job, echoing his involvement in the construction of the Great Western Railway some 20 years earlier, progress on the great ship once again slowed. With over 1,200 men to supervise, and countless drawings and specifications to approve, Brunel came under increasing pressure from Scott Russell's creditors to complete the ship; Adrian Vaughan notes that Martin's Bank, which already owned the yard and was charging the steamship company £2,500 per month in rent, was keen to acquire the ship cheaply if the company should fail. It was therefore imperative that the ship be launched, and launched quickly. Although the stern of the hull had not been completed by June 1857, it was hoped that this would be finished in time for an August launch. This was not to be,

and it was not until the end of October 1857 that all was finally ready.

In the interim, Brunel had once again changed his mind with regard to the method of launching. Although he had originally agreed to slide the ship down the greased wooden slipway on a wooden supporting cradle, this arrangement was abandoned in favour of a series of iron rails which were laid down the launch ramp, with metal strips fixed to the bottom of the carrying-cradles. The contract for the cradles and launching slipways was awarded to the railway contractor Thomas Treadwell; when he was unable to complete the work by the specified date, Brunel and the steamship company were forced to negotiate with Martin's Bank to extend the agreement for renting the shipyard. Not surprisingly, the bank drove a hard bargain, and the need to launch the ship became ever more pressing. Finally, Brunel was forced to announce that the launch would take place on 3 November, the date of the next spring tides.

Since the company's creditors were pressing, Brunel had been unable to spend as much time in preparing and planning the launch as he had hoped. As a result, unlike the successful operations to lift the bridge-trusses at Saltash and Chepstow, the first attempt to launch the ship was an unqualified disaster. There were two distinct parts to the operation, firstly to move the ship down to the river, and secondly to float the ship off her cradle. With regard to the former, it had originally been intended that hydraulic rams would be used to push the ship down the slipway on her wooden supporting-cradle, but this idea was abandoned on the basis of cost; Brunel's son later wrote that it was 'much to be regretted that he did not persist in carrying out his original intention'. The solution then adopted was to use two hydraulic jacks situated at each of the cradles, which were to start the ship moving. In addition, four huge chains ran from the centre of the ship to winches mounted on barges in the middle of the River Thames. There were also chains fixed to both bow and stern, which ran out to other boats moored 300ft out, and back to the shore, where they were attached to a steam 'crab' or winch. As soon as the ship started to move down the slipway, Brunel intended that two enormous brake-drums should be used to ensure that its descent was steady and controlled. In diameter 9ft, and 20ft wide, these drums would pay out huge chains with 60lb links. (It was in front of one of these drums that Brunel would later be photographed by Robert Howlett — an image forever associated with the great engineer, and reproduced on the cover of this book.)

In the days and weeks before the day appointed for the launch, much work had been done, but, while most of the equipment required for the operation had been checked and tested, one vital part, that of the chains and tackle running to and from the barges in the river, had not been satisfactorily

Above: The scene at Millwall, showing the enormous braking drums, and the workmen allocated to operate them. Behind the drum was the platform used by Brunel, his son Henry and his other staff to direct operations on the day of the launch. *Brunel University Library*

MISS HOPE NAMING THE "LEVIATHAN" STEAM-SHIP.

Left: The naming of the ship. Brunel was more concerned about getting the ship safely down the slipway and took little interest in the naming. The name *Leviathan* was soon dropped in favour of *Great Eastern*. *Illustrated London News*

Above: One of the hydraulic jacks used to push the great ship down the slipway. The man with the top hat was Richard Tangye, inventor and supplier of jacks to the project. In all, 21 hydraulic rams were needed to move the 18,000-ton ship down to the water. *Brunel University Library*

Above: The strain is beginning to show on Brunel (far right) in this view taken at the Millwall dockyard during the launch of the *Great Eastern*. Also pictured on the left of the photograph is Brunel's old friend Robert Stephenson, also not a well man. He and Brunel were to die within weeks of each other in 1859. *Brunel University Library*

tested. With creditors pressing, Brunel had little choice but to proceed, since a delay of another month would only add to the company's financial worries. In the early hours of 3 November 1857, the remaining props and supports were removed from the hull, and the slipway rails lubricated with a mixture of oil and black-lead. Brunel had planned that the operation would start two hours before high water, which would be around noon, but long before this it became apparent that, on top of all the engineering problems he might face, he would also have to contend with an enormous audience. Without his knowledge, the hard-pressed directors of the Eastern Steam Navigation Co had sold more than 3,000 tickets for the event, and crowds milled around the shipyard. Brunel himself was reported to have left the platform from which he planned to supervise the operation, to clear crowds from areas of danger. Such was the noise and crush that men stationed around the ship could not see the signals given from the middle of the yard.

Before the ship was to be launched, one final ceremony took place — the ship's naming. This was performed by Miss Hope, the daughter of the company chairman, who broke the obligatory bottle over the ship, christening her *Leviathan*, a name which was later dropped in favour of *Great Eastern*. Brunel had been given a list of names to choose while he was vainly attempting to supervise matters in the packed shipyard; he is said to have retorted that the ship could be called *Tom Thumb*, for all he cared. At almost 12.30, Brunel gave the orders for the braking-chains to be slackened off, and the cradles freed; at first, nothing seemed to be happening, but, after pressure from the river-tackle and the hydraulic rams, the forward cradle began to move, quickly followed by the aft cradle which, accompanied by a loud noise, moved about 9ft. In the course of this, the slack in the braking-chain was taken up rapidly, and the winding-handle of the drum, on which some of the workmen had been leaning, spun round and hit them, throwing some into the air. Five were injured, and one, whose injuries were very serious, died a few days later. Not surprisingly, the scene in the shipyard was chaotic, and operations were suspended to allow Brunel to take stock. Despite heavy rain, another attempt was made later in the afternoon, to no avail, and proceedings were then abandoned. Despite the weather, the crowds still milled around the shipyard, preventing Brunel from investigating why the ship had refused to move very far; in the end, he had little choice but to wait until the following day.

The next morning, having ascertained that little or no damage had been done to the ship, Brunel was faced with the prospect of having to push the ship forcibly down the slipway using hydraulic rams before the next high tide, due in early December. This did not prove easy, and Brunel must have bitterly regretted his decision to abandon his original scheme to use specially-built hydraulic equipment. The river-borne chains and tackle were eventually abandoned as both chains and anchorage failed; throughout November the *Great Eastern* was inched down the slipway, but by the date of the next high tide, 3 December, the great ship was still some distance away from the water. The atmosphere of a siege prevailed, with Brunel spending much time at the yard supervising operations; his already fragile state of health was made much worse by the stress and worry of an operation which seemed never-ending.

In the interminable process of moving the ship there were constant setbacks; the strain of moving such a weight caused not only the river-tackle to give way, but also the hydraulic presses to burst, and their pipes and equipment to fail. It is likely too that some of the problems encountered by Brunel were caused by the unevenness of the slipways, distorted by the great weight of the ship, as well as the skewing of the cradles themselves. Brunel had invited his old friend and rival Robert Stephenson to assist him and give moral support, and it seems likely that it was Stephenson who had encouraged him to increase the number of hydraulic rams used. In a report to directors dated 17 December 1857, Brunel wrote that 'the only mode of proceeding...was to apply considerably more press power'; the successful launching of the *Great Eastern* was eventually dependent on that increased power pushing the cradles down the slip.

By January 1858, 21 hydraulic rams had been made available, either through purchase or loan, including a number from the Birmingham firm of Tangye. When fresh attempts were made to move the ship, early in the new year, there were 10 rams on the forward cradle and 11 on the aft, giving a total force of 4,000 tons, almost double that previously used. When work recommenced on 5 January 1858, considerable progress was made with this increased power, despite the fact that the water in the pipes supplying the hydraulic equipment had frozen. Moving around 5ft during the first day, in the next three the great ship moved another 30ft, so that she was almost in the water. Brunel's son records that by 14 January the *Great Eastern* had been moved almost 200ft down the slipway. Movement was halted until the passing of spring tides lest the ship be floated off unexpectedly, and a week or so later, water was pumped into the ship as ballast to prevent her moving prematurely. It was decided that the floating of the ship should finally take place on Saturday 30 January. Bad weather prevented this, and it was not until the following day that the ballast was pumped out of the hull; as the tide began to rise the *Great Eastern* was finally lifted clear of the cradles and could be floated out into the river.

The euphoria felt by those who had witnessed the struggle to launch Brunel's ship was short-lived. As the ship was towed by tugs to her mooring, debris from the launching-

Above: The only known photograph which shows both Brunel (second from the right) and Scott Russell (far left) together, again at the launch of the *Great Eastern. Brunel University Library*

cradles became entangled with one of the huge paddle-wheels; having cleared this obstruction, the ship then became fouled by a barge, which, despite some fairly energetic encouragement by workmen with axes and sledgehammers, refused to budge. By now tempers and patience were stretched to the limit; it was decided that the easiest expedient would be to sink the barge, and this was duly done. By 19.00 the *Great Eastern* was safely at her moorings off Deptford, and Brunel and the Eastern Steam Navigation Co, whilst no doubt relieved that she was now free, had to count the cost of the launch saga. Originally estimated at £14,000, the launch had in fact cost the company over £120,000. Faced with demands from creditors, still owed £90,000, it was unlikely that the company could raise the money needed to fit out the hull now moored on the River Thames. There the ship remained whilst various schemes for her completion were suggested.

For Brunel, quite apart from the financial loss he himself had sustained in the *débâcle,* the greatest cost was to his health; aside from the sheer physical effort of remaining at the shipyard for long periods, the stress and anxiety he must have endured during those difficult months took their toll, and in May 1858 he and his wife Mary left for France and Switzerland on a recuperative holiday. Brunel was not well enough to return home for Queen Victoria's visit to the ship in June, although most attention appears to have been distracted by the filthy and smelly state of the River Thames, which caused Her Majesty to keep her bouquet pressed into her face for much of the visit.

Returning to London in the autumn, Brunel found that little progress had been made in securing the future of his great ship. There had been considerable debate amongst the directors of the Eastern Steam Navigation Co as to the best way forward, and the board was divided as to whether the ship should be sold, or capital raised to complete the project. It was finally decided that a new company should be formed, and the ship was sold to this body, known as 'The Great Ship Company'. By acquiring the hull of the *Great Eastern* for £165,000 the new venture allowed the old company to be wound up, whilst further capital was raised to complete the ship. Brunel was asked to be the new company's Engineer — a task he willingly undertook, despite the fact that there had been little improvement in his health.

Brunel's poor state of health led to his doctors' ordering him abroad for the winter. Examined by Sir Benjamin Brodie and Dr Bright, he was diagnosed with a kidney ailment, Bright's disease (named after the doctor himself), and left England in late November 1858, not returning until May the following year. Before leaving, he had produced detailed specifications for the fitting-out of the *Great Eastern*, which left out many of the more fanciful or innovative features he had originally hoped to incorporate in his new ship. Two bids were received for the contract: the first, from Wigram &

Below: A painting of the *Great Eastern. National Maritime Museum*

Lucas, valued at £142,000, stuck largely to Brunel's specification, whilst the second, from John Scott Russell, did not. There is little doubt that the directors of the new company were drawn to Scott Russell, who probably knew more about the ship than anyone else apart from Brunel, who was now incapacitated and away from England. The directors were also attracted to the fact that, at £125,000, Scott Russell's tender was considerably lower. Brunel had written to the board in December 1858, stressing the need for Scott Russell to accept the contract on the terms of his specification, something which Russell resolutely refused to do when awarded the contract in January 1859. Work progressed quickly, and was largely complete by July.

In the interim, Brunel and his family had travelled to Egypt, where, after spending Christmas in Cairo, they then travelled up the Nile. Although in Egypt to convalesce, Brunel typically felt well enough to charter a boat to navigate the rapids above Aswân, hardly a relaxing experience! Whilst in Cairo, Brunel had spent Christmas Day with Robert Stephenson, also there to recuperate; it was a chance for the two to meet again after the exertions of the *Great Eastern* launch, which Stephenson had been unable to attend through ill health, despite his staunch support in the weeks and months before. The Brunel family then moved to Italy, spending some time in Rome before returning to England in May 1859, when it was remarked how much better the

engineer looked. This good health was short-lived, and in the last few months of his life, the completion of the *Great Eastern* became the struggle which perhaps sustained him above all else.

Despite his poor health, Brunel visited the ship almost daily, and, to remove the strain of travelling, the family rented a house at Sydenham. There was little doubt that the gaunt engineer now attending to the completion of his ship was a pale shadow of the man who, only two years previously, had posed (rather unwillingly) for Robert Howlett at the Millwall dockyard where the *Great Eastern* then lay. In August, when the ship was almost complete, John Scott Russell held a celebratory banquet on the vessel, but Brunel was too ill to attend. As final preparations were made, it was decided that 7 September would be the date of the ship's first voyage, a trip to the Nore Lighthouse, off Weymouth, to check her compasses. Brunel had already chosen his cabin for the passage, but was destined not to use it. On 5 September Brunel made what would be his final inspection of his ship; sometime after midday he suffered a stroke on board, and was taken back to Duke Street, conscious but terminally ill.

Below: The enormous size of Brunel's *Great Eastern* can be seen in this lithograph published in 1858. At almost 700ft long, she was twice the length of the *Great Britain*. The ship weighed 18,915 tons, as opposed to the 1,340 tons for his first steamship, the *Great Western*. *Illustrated London News*

Above: The *Great Eastern* at anchor after her trial trips. One problem with the great ship was that she was so large that almost all the docks in England were too small to accommodate her. *Illustrated London News*

Right: The *Great Eastern* at Portland, from a drawing by R. P. Lietch. *Illustrated London News*

Brunel would survive for another four days, during which the saga of the *Great Eastern* would take yet another calamitous turn. After a triumphant trip down the River Thames, the ship made her way along the south coast, greeted at every point by enthusiastic crowds, and anchoring off Weymouth on the night of 8 September. The next morning, the ship was steaming back up the English Channel off Hastings, when she was rocked by an enormous explosion, which blew off the forward funnel and sent debris flying into the air. When order had been restored, it was discovered that a feed water-heater had exploded, causing a blowback of boiling water into the engine room. Many of the stokers were badly scalded, and five eventually died from their injuries. The feed water-heater was attached to the two forward funnels of the ship and was a device, also used by Brunel on the *Great Britain*, to warm water on its way to the boilers, by the heat given off from the furnaces below. This had great practical value but on this fateful voyage, stop-cocks placed at the top of each water-heater had been shut off, preventing the escape of surplus steam. As the ship made her way up the Channel, what one writer has described as a 'veritable time-bomb' was set. After the first explosion, a member of the crew opened the valve on the other heater, allowing the steam to escape, and averting a further tragedy.

Brunel, although very ill, had insisted that regular reports of his ship's progress be sent to him, even on his sick-bed. The bad news which then reached him did nothing to improve his worsening condition, and seemed to be the final blow. On the afternoon of 15 September he called his family together, and a few hours later died peacefully. Five days later, in a sombre ceremony at Kensal Green Cemetery attended by his many friends, professional colleagues and a large number of Great Western Railway employees, he was buried in a family tomb alongside his father and mother. The already chequered history of his great ship, the *Great Eastern*, continued; after various mishaps, she was eventually sold at auction for a mere £25,000 and chartered by the Telegraph Construction Co, owned in part by Brunel's friend Daniel Gooch. The ship was then employed to lay the first transatlantic telegraph cable, a use of which her creator would surely have approved.

Below: The *Great Eastern* on the gridiron at Neyland, in Pembrokeshire. The sheer scale of the ship is very apparent in this undated photograph.
Swindon Museum Service

Epilogue
Brunel the Man

In the weeks and months after Brunel's death there were numerous tributes paid by friends, colleagues and the press. *The Times* newspaper, which over the years had often taken the opportunity to snipe at both Isambard and his father, called him an 'eminent engineer...a man who for two generations has been associated with the progress and application of mechanical and engineering science'. He had done the state a great service, the article continued, noting that 'the country was indebted to those who have served her, and their skills which have brought her wealth and strength'. Speaking at a meeting of the Institute of Civil Engineers on 8 November 1859, Joseph Locke addressed the assembled members to record the 'irreparable loss of two most honoured and distinguished members'. This sombre reference to two members alluded to the fact that, within weeks of Brunel's death, his old friend and rival Robert Stephenson had passed away as well. 'In the midst of difficulties of no ordinary kind, with an ardour rarely equalled, and an application of both body and mind almost beyond the limit of endurance Brunel was suddenly struck down', Locke argued, 'before he had accomplished the task his daring genius had set before him'.

Having spent most of this book describing the exploits of this remarkable man, it seems right to attempt to sum him up; both from his own diaries and biographical material published subsequently, it is possible to form some impression of Brunel the man, as well as Brunel the engineer. He was, in all senses of the word, a 'driven' man, and the determination with which he pursued his career at almost any cost, including his health, is testament to this. Despite his outward confidence, which, some argued, bordered on arrogance, Brunel was deeply insecure, and his diaries reveal that he worried long and hard that all his 'castles in the air' would come crashing down. One obvious result of this drive and ambition was the long hours he worked and his frequent absences from home. This meant that he often spent less time with his family than he would have liked, and it was left to his wife Mary to provide a stable home environment for his children, although he appears to have had a close relationship with them when he could be at home.

Although he built on a grand scale, he was also a perfectionist, and Brunel's skills as a draughtsman and engineer meant that his standards were of the highest order, whether it be in relation to building materials, workmanship or conduct. This trait, although ensuring that the work he produced was of the best, could also cause problems, since he often changed his mind and made minor alterations as projects proceeded. This mercurial behaviour did not always make him popular with contractors or shareholders, as was the case both on the South Devon Railway and the *Great Eastern*.

It may also be apparent, from some of the events retold in this book, that Brunel was not the easiest man to work with or for. Brunel refused in general to work in partnership with any other engineer. Writing in 1836, he explained that to do so might mean that any project could 'be seriously embarrassed by differing opinions'. He appears to have largely maintained this policy, but in the case of the *Great*

Above: Drawing of Brunel. *Swindon Museum Service*

Eastern, he ignored his own advice, working with John Scott Russell as a partner, and the resultant disastrous turn of events appears to justify his original argument. Brunel expected much from the staff and contractors he employed, who received short shrift when work or behaviour were not up to standard. In October 1840 he wrote to one of his staff, describing him as 'a cursed lazy inattentive apathetic vagabond'; this unfortunate assistant had offended his master's sensibilities by making drawings on the backs of others, and by his 'criminal laziness'. Contractors often fared little better, and found Brunel a hard taskmaster, checking and altering specifications, and sometimes taking a long time to authorise payment for work done, a situation which drove some close to bankruptcy.

Despite the public acclaim Brunel received for a good deal of his work, throughout his career he also came in for an equal amount of criticism, much of it for what shareholders and newspapers regarded as extravagance. Many of the projects in which he was involved appear to have had hopelessly optimistic budgets, and this, coupled with the fact that Brunel generally used the best materials and methods available, meant that few projects were completed within budget, with even his beloved Great Western Railway costing £6.5 million, almost double his original estimate. The broad-gauge, the adoption of the atmospheric system in south Devon and the *Great Eastern* all involved expensive errors of judgment which were to cost the companies involved dear, although as Daniel Gooch, Brunel's friend, remarked of him in his diary, 'great things are not done by those who sit down and count the cost of every thought and act'.

The death of both Brunel and Stephenson in 1859 was almost the end of an era, since the pace and emphasis of change in 19th-century Britain was subtly changing. When Brunel had begun his career, Britain was still largely a rural country, with a population of around 14 million; by the death of Queen Victoria it had become an industrialised nation of over 30 million. The railways planned and laid out by Brunel and his contemporaries had encouraged the growth of urban development, and the large-scale industry which accompanied it, and brought the countryside within easy reach of cities for the first time. By the 1850s the role of the engineer himself was also changing, and there was less of a pioneering spirit prevalent. Brunel, in his early years, had been able to move easily between the marine, civil, railway and mechanical engineering fields, a situation which later engineers were not able to do, as the profession began to divide into more specialist fields.

The small Brunel family grave in Kensal Green Cemetery is easy to miss in a graveyard full of grand Victorian tombs; a rather grander memorial, a striking statue of Isambard by the sculptor Carlo Marochetti, was commissioned by his family and friends, and now stands on the Embankment in

Above: Portrait of Brunel. *Swindon Museum Service*

London. A stained-glass memorial window was also placed in the nave of Westminster Abbey by the Brunel family in 1863. Perhaps the most telling memorials to the man are the large number of his structures which have survived to the present day; most of his Great Western main line still exists, and, at the time of writing, its significance has been marked by its nomination as a potential World Heritage Site by UNESCO, an accolade of which Brunel would no doubt have approved. In the West of England, although his timber viaducts have long since been replaced, the Royal Albert Bridge still spans the Tamar, carrying his railway into Cornwall. Fittingly, in Bristol, the city where his career first blossomed, there is much to see, including the SS *Great Britain*, now back in the dockyard where she was built. Above the Avon Gorge, the Clifton Suspension Bridge, built by Brunel's contemporaries as a memorial to him, is surely a fitting tribute in itself. The great engineering achievements chronicled in this book, coupled with the familiar image of the man himself — the stovepipe hat, the crumpled clothes and the cigar — will ensure that Isambard Kingdom Brunel will continue to be remembered as one of the most significant figures of the 19th century and beyond.

Chronology

1806   Brunel born in Portsea, Portsmouth

1820   Sent to France to study at Caen and Paris

1822   Begins work for his father Marc Isambard Brunel

1825   Work on Thames Tunnel commences

1827   Brunel appointed Resident Engineer on Tunnel
Tunnel flooded: work restarts after six months of
repairs

1828   Work on Tunnel abandoned after second major flood;
Brunel badly injured

1830   Wins competition to design Clifton Bridge at Bristol

1833   Appointed as Engineer to the Great Western Railway

1835   Great Western Railway Bill passed by Parliament

1836   Work begins on Clifton Bridge
Marries Mary Horsley
Keel of SS *Great Western* laid

1837   SS *Great Western* launched

1838   London-Maidenhead section of GWR opens

1839   Work on SS *Great Britain* begins
Completion of Maidenhead Bridge

1841   The Great Western Railway opens fully from London
to Bristol

1842   Opening of first section of Bristol & Exeter Railway

1843   SS *Great Britain* launched at Bristol
Thames Tunnel finally opens to the public
Swindon Works opens

1844   Appointed as Engineer to the South Wales Railway
Completion of Bristol & Exeter Railway

1845   Opening of Hungerford Suspension Bridge, London
Maiden voyage of SS *Great Britain*

1846   First section of Atmospheric Railway opens on
South Devon line
SS *Great Britain* runs aground in Dundrum Bay

1847   SS *Great Britain* salvaged

1848   The 'Atmospheric Caper' abandoned on the South
Devon Railway

1849   Isambard's father, Sir Marc Isambard Brunel dies,
aged 81

1852   Completion of Chepstow Bridge on South Wales line
Opening of West Cornwall Railway

1853   Commencement of work on Royal Albert Bridge
at Saltash

1854   Opening of new Paddington station
Work begins on SS *Great Eastern*

1855   Designs for prefabricated hospital for Crimea produced

1857   First aborted attempt to launch SS *Great Eastern*
Great Western scrapped

1858   SS *Great Eastern* finally launched

1859   Royal Albert Bridge opened by Prince Consort
Boiler explosion on SS *Great Eastern* during trials
Brunel dies, aged 53

1860   Maiden transatlantic voyage of SS *Great Eastern*

1864   Opening of Clifton Suspension Bridge

1892   Brunel's broad-gauge finally abolished

Sources & Bibliogrpahy

Something of the importance attached to the contribution made by Isambard Kingdom Brunel can be gathered from the huge number of books and articles written about him and the various aspects of his work. As well as the primary material in the collections of Bristol University Library and the Public Records Office, *The Times* newspaper and the *Illustrated London News*, the author has drawn on the works of those writers and biographers who have already shone a light on the man they called the 'Little Giant'. *The Life of Isambard Kingdom Brunel*, written and compiled by his son Isambard in 1870, is an excellent source of information and deals comprehensively with his career, particularly since many of the contributors had worked for, or with, Brunel himself. However, the book fails to shed much light on the man rather than the engineer. Lady Celia Noble's account of both Marc and Isambard, published in 1938, more than makes up for this deficiency, as do the two seminal biographies of Isambard written by L. T. C. Rolt in 1957 and Adrian Vaughan in 1991. Published 34 years apart, the latter present a fascinating contrast in styles, but bring the story to life none the less. Peter Hay's account of Brunel, published in 1973, is also an excellent account of both the man and the historical context within which he worked. Other sources consulted are listed below under the chapters of the book to which they relate. I only hope that, in such esteemed company, my own effort does justice to the story.

General Works

Beamish, R., *Life and Memoir of Sir Marc Isambard Brunel*, Longman, 1862

Beckett, D., *Brunel's Britain*, David & Charles, 1988

Bentley, N., *The Victorian Scene*, Spring Books, 1971

Brunel, Isambard, *The Life of Isambard Kingdom Brunel*, Longman, 1870

Casson, H., *Victorian Architecture*, Art & Technics, 1948

Gloag, J., *Victorian Comfort: A Social History of Design 1830-1900*, David & Charles, 1973

Hay, P., *Brunel: Engineering Giant*, Batsford, 1985

Lambton, L., *Vanishing Victoriana*, Phaidon, 1976

Minchinton, W., 'I. K. Brunel, Engineer 1859', *History Today* Vol 29 December 1979, pp824-831

Newhouse, F., 'Isambard Kingdom Brunel: A Reappraisal', *The Engineer*, 11 September 1959

Powell, R., *Brunel's Kingdom*, Watershed Media Centre, 1985

Pugsley, A., *The Works of Isambard Kingdom Brunel*, Institute of Civil Engineers and University of Bristol, 1976

Quartermaine, A., 'I. K. Brunel — The Man and His Works', *Proceedings of the British Railways (Western Region) London Lecture & Debating Society*, 5 March 1959

Rolt, L. T. C., *Isambard Kingdom Brunel*, Pelican Books, 1982

Simmons, J., *Victorian Railways*, Thames & Hudson, 1995

Vaughan, A., *Isambard Kingdom Brunel: Engineering Knight-Errant*, John Murray, 1991

1. Tunnels, Bridges and Docks

Baker, J., (Ed), *The New Guide to Bristol & Clifton*, Baker & Son, 1898

Barlow, W. H., 'A Description of the Clifton Suspension Bridge', *Proceedings of the Institute of Civil Engineers*, Vol 26, 1867, pp243-257

Buchanan, R. A., 'I. K. Brunel and The Port of Bristol', *Transactions of the Newcomen Society*, December 1969, pp41-56

Buchanan, R. A. and Williams, M., *Brunel's Bristol*, Redcliffe Press, 1982

Cotterell, A. E., *The History of the Clifton Suspension Bridge*, Clarks Printing Services, 1928

Falk, N., *Brunel's Tunnel and Where It Led*, Brunel Exhibition Project, 1980

Greenacre, F. and Stoddard, S., *The Bristol Landscape: The Watercolours of Samuel Jackson*, City of Bristol Council, 1986

Lucas, W. T., *Notes on Old Clifton*, published by the author, 1930

McIlwain, J., *Clifton Suspension Bridge*, Pitkin Guides, 1996

Thames Tunnel Company, *An Explanation of the Works of the Tunnel Under the Thames*, 1846

2. The Birth of the Great Western Railway

'Brunel's Broad Gauge', *The Locomotive Magazine*, 15 June 1942, pp104-106

Bryan, T., *North Star: A Tale of Two Locomotives*, Thamesdown Borough Council Museums Service, 1989

Channon, G., *Bristol and the Promotion of the Great Western Railway*, Historical Association (Bristol Branch), 1985

Francis, J. F., *History of the English Railway 1820-1845*, Longman, 1850

McDermot, *History of the Great Western Railway, Vol 1: 1833-1863*, Great Western Railway, 1927

Measom, George, *Official Illustrated Guide to the Great Western Railway*, Richard Griffin & Co, 1852

Reports of Half-Yearly Meetings of the Great Western Railway Company, (GWR Museum collection)

Simmons, J., (Ed), *The Birth of the Great Western Railway: Extracts from the Diary and Correspondence of George Henry Gibbs*, Adams & Dart, 1971

The Railway Gazette, GWR Centenary Supplement, 30 August 1935

Wilson, R. B., (Ed), *Sir Daniel Gooch: Memoirs and Diary*, David & Charles, 1972

3. Brunel's Railway

Arthurton, A., 'The Older Engineering Works of the Great Western Railway', *Great Western Railway Magazine*, August 1907, pp176-177

Atkins, P., 'Box Railway Tunnel and I. K. Brunel's Birthday: A Theoretical Investigation', *Journal of British Astronomical Association*, Vol 95, June 1985, pp260-262

Biddle, G. and Nock, O. S., *The Railway Heritage of Britain*, Michael Joseph, 1983

Bourne, J. C., *History and Description of the Great Western Railway*, David Bogue, 1846

Bristol and its Association with the GWR 1835-1935, Great Western Railway, 1935

Cattell, J. and Falconer, K., *Swindon: The Legacy of a Railway Town*, HMSO, 1995

Chapman, W. G., *Track Topics*, Great Western Railway, 1935

Cockbill, T., *Finest Thing Out*, Quill Press, 1988

Cooke, R. A., *Atlas of the Great Western Railway*, Wild Swan Books, 1988

Gooch, D., *Diaries of Sir Daniel Gooch*, Kegan Paul, 1892

'Great Railway Tunnels, No 1: Box Tunnel', *Great Western Railway Magazine*, September 1928, pp356-358

Murray, J. F., *A Picturesque Tour of the River Thames in its Western Course*, H. G. Bohn, 1853

South, R., *Crown, College & Railways*, Barracuda Books, 1978

Vaughan, A., *A Pictorial Record of Great Western Architecture*, OPC, 1977

Waters, L., *Rail Centres: Reading*, Ian Allan Publishing, 1990

White, H. P., *Regional History of the Railways of Great Britain, Vol 3: Greater London*, David & Charles, 1971

4. The Bristol Steamships

Buchanan, R. A., 'I. K. Brunel and The Port of Bristol', *Transactions of the Newcomen Society*, December 1969, pp41-56

Claxton, Christopher, *A Description of the* Great Britain *Steamship*, John Taylor, 1845

Conservation Plan for the Great Western Steamship Company Dockyard and SS Great Britain, SS *Great Britain* Project, 1999

Corlett, E., *The Iron Ship*, Conway Press, 1990

Dumpleton, B. and Miller, M., *Brunel's Three Ships*, Venton, 1974

Farr, G., *The Steamship* Great Britain, Historical Association, 1965

Farr, G., *The Steamship* Great Western, Historical Association, 1965

Fogg, N., *SS* Great Britain: *Brunel's Flagship of the Steam Revolution*, SS *Great Britain* Project, 1996

Gibbs, G. E., *Passenger Liners of the Western Oceans*, Staples Press, 1952

Griffiths, D., *Brunel's* Great Western, PSL Books, 1985

Harris, M., (Ed), *Brunel, the GWR & Bristol*, Ian Allan Publishing, 1985

Latimer, T., *Annals of Bristol in the Nineteenth Century*, Kingsmead Press Reprint, 1970

5. The Broad-Gauge Empire

The South Devon Atmospheric Railway, Broad Gauge Society, 1998

Carpmael, R., 'Brunel: The First Engineer of the Great Western Railway Company', *Proceedings of the Great Western Railway (London) Lecture & Debating Society*, 20 October 1932

Chapman, W. G., *Track Topics*, Great Western Railway, 1935

Chepstow Society, *Brunel's Tubular Suspension Bridge Over the River Wye*, Ivor Waters, 1970

Clayton, H., *Atmospheric Railways*, F. H. Clayton, 1966

Day, L., *Broad Gauge*, Science Museum/HMSO, 1985

Great Western Railway, *Report of a Special General Meeting*, 19 January 1843

Gregory, R. H., *The South Devon Railway*, Oakwood Press, 1982

Hadfield, C., *Atmospheric Railways*, David & Charles, 1967

Minutes of Evidence Taken Before the Commissioners Appointed to Inquire into the Gauge of Railways, HMSO, 1846

Kay, P., *Exeter-Newton Abbot: A Railway History*, Platform 5 Books, 1993

Maggs, Colin, *Rail Centres: Bristol*, Ian Allan Publishing, 1981

Maggs, Colin, *The Swindon to Gloucester Line*, Alan Sutton, 1991

Prospectus, Oxford, Worcester & Wolverhampton Railway, September 1844

'The River Wye Bridge: Centenary of the Chepstow Structure', *British Railways Magazine*, Vol 3 No 9, 1952, pp169-170

'£. s. d.', *The Broad Gauge: The Bane of the Great Western Railway Company*, John Olliver, 1846

6. Saltash and the *Great Eastern*

The Broad Gauge in Cornwall, Broad Gauge Society, BGS, 1996

'Brunel's Timber Viaducts', *Railway Gazette Supplement*, 30 August 1935, pp47-49

Chapman, W. G., 'Saltash Bridge: an Octogenarian', *Great Western Railway Magazine*, May 1939, pp 207-209

Dumpleton, B., 'Brunel's *Great Eastern*', *World Models*, January 1978, pp28-30

Emmerson, G., *SS* Great Eastern: *The Greatest Iron Ship*, David & Charles

Osler, E., *History of the Cornwall Railway 1835-1846*, Avon Anglia Publications, 1982

Woodfin, R. J., *The Centenary of the Cornwall Railway*, Jefferson & Son, 1960

Index

Cover, main image: The truly graceful lines of the Maidenhead Bridge are revealed in this Victorian lithograph. *Swindon Museum Service*

Front cover, left inset: SS *Great Britain. National Maritime Museum*

Front cover, right inset: Portrait of I. K. Brunel. *Playback Productions*

Back cover, left inset: A J. C. Bourne lithograph of a broad-gauge train. *Author's Collection*

Back cover, right inset: Clifton Suspension Bridge. *Author's collection*

Below: The atmospheric pumping house at Starcross, long after the removal of the system. For some time a small museum dedicated to the railway was situated in the building, but it is now occupied by a yacht club. *Swindon Museum Service*